W9-CRB-289

Safety in Chemical Production

International Union of Pure and Applied Chemistry

Safety in Chemical Production

Proceedings of the First IUPAC Workshop on
Safety in Chemical Production
Basle, Switzerland, 9–13 September 1990
Edited by J. Whiston

Group Safety Manager, ICI, London

OXFORD

BLACKWELL SCIENTIFIC PUBLICATIONS

LONDON EDINBURGH BOSTON
MELBOURNE PARIS BERLIN VIENNA

CHEMISTRY

4782811

© 1991 International Union of Pure and
Applied Chemistry and published for them by
Blackwell Scientific Publications
Editorial offices:
Osney Mead, Oxford OX2 0EL
25 John Street, London WC1N 2BL
23 Ainslie Place, Edinburgh EH3 6AJ
3 Cambridge Center, Cambridge
  Massachusetts 02142, USA
54 University Street, Carlton
  Victoria 3053, Australia

Other Editorial Offices:
Arnette SA
2, rue Casimir-Delavigne
75006 Paris
France

Blackwell Wissenschaft
Meinekestrasse 4
D-1000 Berlin 15
Germany

Blackwell MZV
Feldgasse 13
A-1238 Wien
Austria

All rights reserved. No part of this publication may be
reproduced, stored in a retrieval system, or transmitted,
in any form or by any means, electronic, mechanical,
photocopying, recording or otherwise without the prior
permission of the copyright owner.

First published 1991

Printed in Great Britain at
the University Press, Cambridge

DISTRIBUTORS

Marston Book Services Ltd
PO Box 87
Oxford OX2 0DT
(Orders: Tel: 0865 791155
        Fax: 0865 791927
        Telex: 837515)

Australia
Blackwell Scientific Publications
(Australia) Pty Ltd
54 University Street
Carlton, Victoria 3053
(Orders: Tel: 03 347-0300)

Distributed in the USA and North America by
CRC Press, Inc.
2000 Corporate Blvd, NW
Boca Raton
Florida 33431

British Library
Cataloguing in Publication Data
IUPAC Workshop on Safety in Chemical Production
  (1st : 1990 : Basle, Switzerland)
Safety in chemical production.
I. Title II. Whiston, J. III. International
Union of Pure and Applied Chemistry
660.2

ISBN 0632032553

Library of Congress
Cataloguing in Publication Data
IUPAC Workshop on Safety in Chemical Production
  (1st : 1990 : Basel, Switzerland)
    Safety in chemical production : proceedings of the
  First IUPAC Workshop on Safety in Chemical
  Production, Basle, Switzerland, 9–13 September
  1990 : edited by J. Whiston.
        p. cm.
    At head of title: International Union of Pure and
  Applied Chemistry.
    ISBN 0-632-03255-3
    1. Chemical Industry—Safety measures—
  Congresses. I. Whiston, J. II. International Union
  of Pure and Applied Chemistry. III. Title.
  TP149.I85   1990
  660'.2804—dc20

Contents

Day 2: Case Studies

Day 3: Safety Education Today

Organizing committee

W. M. CORCORAN *Vice President for Public Affairs, Allied–Signal, Morristown, USA*

L. CORIGLIANO *Head of Safety and Environmental Protection, Montedipe, Milano, Italy*

J. B. DONNET *CNRS University of Haute Alsace—ENSCM, Liaison to Chemrawn Committee of IUPAC, Mulhouse, France*

K. EIGENMANN *Head of Central Safety Services, Ciba–Geigy AG, Basel, Switzerland*

A. FISCHLI *COCI, IUPAC, Secretary, Basel, Switzerland*

R. FORLEN *Convention Services, Ciba–Geigy AG, Basel, Switzerland*

E. J. GRZYWA *Instytut Chemii Przemyslowej, Warszawa, Poland*

E. GWINNER *Public Relations Group, Hoffmann–La Roche AG, Basel, Switzerland*

H. KUNZI *Head of Safety and Environmental Protection, Hoffmann–La Roche AG, Basel, Switzerland*

H. J. PASMANS *TNO Prins Maurits Laboratory, Liaison to European Federation of Chem. Engineering, Rijswijk, The Netherlands*

V. PILZ *Managing Director, Plant Safety and Technical Inspection, Bayer AG, Leverkusen, Germany*

W. REGENASS *Deputy Director Pigments, Ciba-Geigy AG, Basel, Switzerland*

L. REH *Institute for Process and Cryogenic Engineering, Swiss Federal Institute of Technology, Zurich, Switzerland*

W. RICHARZ *Institute for Chemical Engineering + Industrial Chemistry, Swiss Federal Institute of Technology, Zurich, Switzerland*

W. RISER *Convention Service, Ciba–Geigy AG, Basel, Switzerland*

J. J. SALZMANN *Head of Corporate Safety and Environmental Protection, Sandoz, Basel, Switzerland*

H. WEGMULLER *Research and Development Dyestuffs and Chemicals Division, Ciba–Geigy AG, Basel, Switzerland*

J. WHISTON *Group Safety Manager, ICI, London, United Kingdom*

F. WIDMER *Vice President (planning/development) of Swiss Federal Institute of Technology (ETH Zurich), Institute for Process and Chemical Engineering, Zurich, Switzerland*

D. WYRSCH *COCI, IUPAC, Chairman, Basel, Switzerland*

Sponsors

The organisation of the conference was made possible thanks to the generous support of the following companies and organisations (this assistance is once again gratefully acknowledged):

Akzo NV, *Netherlands*

Alusuisse–Lonza Holding AG, *Switzerland*

BASF AG, *Germany*

Bayer AG, *Germany*

Boehringer Ingelheim, *Germany*

Ciba–Geigy AG, *Switzerland*

Degussa AG, *Germany*

Dow Europe SA, *Switzerland*

Exxon Chemical International Inc, *Belgium*

Gerling–Schaden-Institut, *Germany*

Henkel KGaA, *Germany*

Hoechst AG, *Germany*

Hoffmann–La Roche AG, *Switzerland*

Huels AG, *Germany*

ICI Plc, *United Kingdom*

Montecatini Edison SPA, *Italy*

National Versicherung, *Switzerland*

Nestel SA, *Switzerland*

Nobel Industries, *Sweden*

Rhone–Poulenc SA, *France*

Roussel–Uclaf, *France*

Sandoz, *Switzerland*

Schering AG, *Germany*

Shell International Chemical Company Ltd, *United Kingdom*

Solvay & Cie SA, *Belgium*

SUVA, *Switzerland*

Swiss Academy of Science, *Switzerland*

Swissair, *Switzerland*

The Association for the Progress of New Chemistry, *Japan*

The Royal Society, *United Kingdom*

The Wellcome Foundation, *United Kingdom*

Toray Science Foundation, *Japan*

UNEP, *Kenya*

UNIDO, *Austria*

Winterthur Safety Engineering, *Switzerland*

Zürich Versicherungsgruppe, *Switzerland*

Preface

This first IUPAC Workshop has its origin in the General Assembly discussion in Lyon in 1985, held at a time when there were significant concerns following the events at Seveso and Bhopal, in particular on the procedures and processes used when safety technology was transferred from the developed to the developing world. It was also recognised that there were many groups active in the fields of safety education; thus it was essential to identify the additive and distinctive contributions which IUPAC could make to this area.

After detailed discussion three points were clearly evident:
- IUPAC is the only truly international organisation in Chemistry.
- IUPAC is one of the few organisations independent of both Government and Industry. It is uniquely placed in bringing academic scientists together with technical personnel from Industry and Government—this has been one of its many roles over its history.
- IUPAC can influence the science teaching in Universities and hence the education in safety technology of Chemists and Chemical Engineers.

The IUPAC Committee on Chemistry and Industry (COCI) recognised that University teaching in this area was inadequate. Thus if a Workshop could be designed which would link together real case studies, especially in the field of Hazard Analysis and Risk Reduction, with model lectures based on the case studies for use in Universities then a real need would be met.

It was also determined to use an industry base to demonstrate the link between the University science and its application in Industry. Basle was chosen for practical reasons: three multi-national Chemical Companies are located conveniently near to each other and they agreed to offer the facilities to the Workshop.

Day one of the Workshop provides the scientific basis for Risk and Hazard Analysis. The application of these principles is seen in the detailed case studies provided by the Basle Chemical Industry on Day two. With a scientific base and an appreciation of the practical applications then on Day three the way in which principles and practices can be brought into a University Curriculum is demonstrated including some typical model lectures. Finally on Day four views on principles, practices and teaching are collected, key principles established and suggestions made for future development.

The Workshop proceedings include not only papers presented but also the 'model' lectures, a bibliography of possible teaching aids as well as two papers which would have been included in Day one presentations had additional time been available.

Final decisions for follow-up have yet to be reached but if the Workshop is to be of continuing value then it must be only the first of many.

J. Whiston

Opening remarks

Dr D. Wyrsch (Chairman of the Standing IUPAC Committee on Chemistry and Industry) gave a general welcome. He explained that the idea for the Workshop had been stimulated by the Bhopal accident and had been organised by a very active international committee. He thanked the members of that committee and other supporters and financial contributors. An aim had been to have a wide attendance to discuss current practice in safety assessment and risk reduction and how to encourage education programmes in Chemical Engineering and in Chemistry curricula. A wide attendance had been achieved with 44 countries represented (both developed and developing) and a mix from academia and industry (of 134 participants, 56 were from Universities, 56 from industry and 7 from Government).

He hoped the workshop would be a success. If so, others would follow.

Professor Y. P. Jeaninn (President of IUPAC) then summarised the aims and origins of IUPAC. He noted that it had originally been founded by industrial chemists, but initially it was mainly Government and Academia agencies that were involved. In order to stimulate industry involvement the idea of industrial associates was developed and now there are 177. Then in 1977 a special committee for the chemical industry was set up to advise on problems to be tackled. The idea for the Workshop came from that committee.

He commented that the chemical industry is a fact of life, it is part of the environment. We are all chemical users and have to accept that there are problems. But these problems have to be addressed, globally, both by industry and others. Keeping the environment clean is everybody's wish, both producers and users, but risk is a fact of life also. There cannot be zero risk, but it is right to reduce risk as far as is practicable.

He wished the workshop success giving special thanks to Ciba–Geigy, Hoffman–La Roche and Sadoz for their support.

Mr E. Keller (Member of the Basle Government) gave a welcome on behalf of Basle. He noted Basle's strong links with IUPAC. He noted also Basle's first hand experience of chemical safety problems; some had been bitter, but had resulted in improvements. He hoped the workshop would stimulate further progress.

He commented that knowledge is essential, but is not in itself enough; it has to be promulgated and used. Training is vital and he thought Basle industry has a major part to play. The human factor is a cause for concern and chemical safety has become a political issue. The sheer size of plants and the size of organisations seems to be a source of worry to the lay public. Industry has to work at restoring and maintaining the trust of the public.

He wished the workshop would be successful as a means of sharing experience, avoiding barriers and stimulating open discussion.

D. Wyrsch, Y. P. Jeannin and E. Keller

Day 1: Principles of Safety Assessment and Risk Reduction

General introduction and survey of programme

Prof Dr F Widmer

Vice-President Planning/Development, ETH, Zurich

It is for me a great honour to act as chairman of the first day of our workshop and to generally introduce you by some words into the programme.

Safety has always been a basic need of human being. It was the **nature** against which human being first had to protect himself. Illness, cold, dryness, heat, flood, earthquakes, animals and finally human being himself threatened him.

Thus, human being had to look after his own safety since his creation. He had to learn how to deal with dangers - and he learned from experience, as he still does today.

As time passed by, mankind learned how to use technology. Consequently, the dangers which occured by applying these new technologies came to the fore. Today, the dangers of technology are the centre of discussion and dispel, in public, the discussions about the natural dangers with which we are confronted. People often forget to mention that parallel to the development of technologies and sciences the methods to control the arising dangers have considerably improved, too.

In this connection, Richard von Weizsäcker recently formulated the following (translated faithfully):

> **"Our welfare can only be guaranteed if we approve and participate in the technological progress. For that, however, we have to understand it, which also means to get to know and to master its dangers and risks."**

There is no doubt that safety has to be an inseparable, integral part of research and development as well as of education. First of all, this is directed at universities and all other places of training. Within this workshop, we want to treat and discuss - out of the large subject "safety in chemical production" - the better understanding of the principles of safety assessment and risk reduction and, above all, the education in safety. Here in Basel, we have got the unique opportunity to profit from the high state of art in safety and from the broad experience in safety technology of the local process industry.

If we generally analyse the causes of accidents, they often are a result of two or more simple failures and deviations among components of an installation which interact in a certain unexpected and undesired way. This interacting tendency is a characteristic of most systems, especially of bigger plants in process industry. Safety measures additionally contribute to a higher level of complexity and thus, interactions. Therefore, we have to pay special attention to the interaction of deviations.

Let me illustrate this phenomenon by an accident which happened about half a year ago in our country:

An electrically heated, large boiler, providing hot water for some installations of a community, burst, completely destroying the house where it was installed - including a flat - and killing a young family of three.

How could this fatal accident happen? Some days before the accident, the janitor recognized the overtemperature of the boiler and immediately informed the responsible community administration. An electrician was ordered to switch off the electrical current in order to stop a further heating of the water. Despite this emergency action, the janitor recognized the following day an even further increase of the temperature and of the pressure in the boiler. Again, he informed the community administration, but unfortunately, the electrician was not available this day. In the evening of the same day, the boiler burst due to overpressure despite the existing emergency relief system. According to an information recently published in the newspapers, the investigations of this fatal accident showed that, firstly, the electrical current of the heating was obviously not switched off because of an inexplicable failure of the switch, and, secondly, that the installed emergency relief valve was wrongly designed and did not work.

If we arise the question: "What happened really and what was the primary cause of this accident, being the result of totally 5 failures and deviations?"

1) Was it **human error** by the community administrators who did not look for an electrician at all costs?

2) Was it a **technical failure**, whether of the temperature control or of the switch or of the emergency relief valve?

3) Or was it an **unsuitable** or wrong **design of the system** to place a flat above a boiler installation?

If you answered "not sure" or "no" to all of the above questions, I agree with you. The cause of the accident is to be found in the interaction of the system. Only one of these failures or deviations prevented, and the accident or the fatal consequences could have been avoided.

Indeed, it is in a certain case the **interaction of the multiple failures** that explains the accident. The failures only become serious when they interact. Thus, the more complex a system is - as in the case of most of today's plants of the process industry - the bigger is the danger that, by interactions of deviations, serious consequences can occur.

The key to solve these extensive and complex safety problems is a formal and systematic risk analysis of a process plant.

This was one of the reasons why we chose the procedure of risk analysis as frame for the presentations of the first day of our workshop. Furthermore, the risk analysis and hazard analysis are interesting fields included in university training programmes, because they illustrate the applications of scientific principles.

The procedure of risk analysis - applied in the process industry in Switzerland - is roughly consisting of the following steps:

Collection of basic data like hazardous properties including toxicity, thermal stability, flammability and explosivity.

Definition of safe process conditions and potentially hazardous deviations of it.

Identification of hazards by means of systematic procedures, with regard to human error, technical failures, procedure used, environment as well as to the design of system.

Hazardous analysis and risk evaluation, characterized by the technical analysis of the severity of possible consequences and the probability of occurence of an accident caused by the hazards.

Risk. If the evaluated risk exceeds the level of the acceptable risk or is considered as too high, additional

Safety measures are required for reducing the possible consequences and/or the probability of occurence of an accident. Technical, organisational and personnel safety measures are taken into consideration.

Each step is connected with a loop with previous steps, for instant, technical safety measures will lead to extended hazard identification, and so on. These procedures will be continued as long as the level of the minimised or acceptable remaining risk is reached.

The scientific programme of our workshop is built up in such a way that the presentation of the first day, following the procedure of the risk analysis, serves as basic scientific and methodical preparation for the detailed case studies in 12 different groups of tomorrow.

The first two days together, on the other hand, should provide you fundamentals, applications and impressions for the third day, that means for the discussion of <u>safety education today</u> and for the discussion of the <u>case studies to be used in university training programmes</u>.

Fact-finding and basic data
Part I: Hazardous properties of substances

Richard L. Rogers

ICI Fine Chemicals Manufacturing Organisation, Manchester, UK

Abstract - Safety in chemical production requires a knowledge and understanding of the hazardous properties of the materials being handled. In contrast to physical chemical properties, hazard data is not an inherent characteristic of a substance, rather it depends on and is changed by the interaction of the substance with the plant situation. Thermal explosions resulting from the thermal instability of chemicals only occur if the rate of heat loss from the system is less than the rate of heat production. This influences not only the experimental methods used to measure these properties but also the interpretation of the data obtained and its relationship to plant scale operations.

INTRODUCTION

The origin of the safety problem which arises not only in the manufacture of chemicals but also in their transport, storage and use is the inherent hazard potential hidden within a substance. However it is critically important to recognise that the possession of hazard potential by a substance does not make that chemical hazardous per se. For a hazardous situation to occur i.e. the release of a chemical's hidden potential, a trigger mechanism must also be present. This trigger or activation potential resides in the plant/equipment, chemical process or method of handling of the chemical.

For example, water is a necessity for life, however it could also be considered as toxic since it can cause drowning. Similarly wood in a tree is difficult to ignite and burns relatively slowly but when finely divided, as occurs in a sawmill, the resultant fine wood dust can be readily ignited and lead to a violent dust cloud explosion.

It is evident therefore, that the hazard characteristics of a material cannot be simply considered, measured and tabulated as other physico-chemical properties of substances such as melting point, boiling point etc. Hazardous properties vary with physical form and depend markedly on the interaction of a material with the associated equipment or usage.

In evaluating hazards it is therefore necessary to obtain data on both the substances being handled and the equipment or process within which they are being used.

The principle hazards of substances arise from their toxicity, ecotoxicity, explosivity, chemical reactivity and flammability. In addition, information on a material's physical chemical properties such as vapour pressure, volatility, acidity, etc. are also required before a hazard assessment or analysis can be carried out.

Toxicity and ecotoxicity are a measure of the harmful effect of a substance on man and his environment. Data on these are required both to enable decisions on the degree of containment or allowable exposure to be made at the design stage and during operation and also to evaluate the possible consequences if an incident was to occur.

If triggered, the explosivity, chemical reactivity and flammability of a substance, can cause a hazardous situation to occur within the plant units themselves and in addition may lead to a release of a chemical to the surroundings with the attendant toxic and ecotoxic problems.

This paper concentrates on the identification and evaluation of hazardous properties

of single substances and their undesired decompositions. The hazards associated with
carrying out desired reactions in chemical manufacture are covered in Part II while the
problems associated with flammability and ignition risks are considered in Part III.

EXPLOSIONS

An explosion can be considered as a sudden release of energy. This is usually accompanied
by a rapid increase in pressure due to either the production of gas or the physical effect
of increased temperature on the components present. Traditional detonations or
deflagrations occur when a combustion wave travels through a flammable gas, vapour or dust
air mixture or after the initiation of a "true" explosive. In such cases the energy is
instantaneously released once the system has been initiated by an external stimuli.

In contrast thermal explosions caused by runaway chemical reactions are invariably
preceded by a self heating process. Thus if the rate of heat release by the chemical
reaction is greater than the rate of heat loss to the surroundings a temperature rise will
occur which leads to acceleration of the exothermic reaction.

The difference in initiation and development true and thermal explosions is important in
the characterisation of materials. However although the course of true and thermal
explosions are different, the consequences in burst reactors or storages and on the plant
and its environment are similar.

RUNAWAY REACTIONS AND THERMAL STABILITY

The occurrence of a runaway reaction or thermal explosion depends not only on the rate of
heat generation from a chemical reaction but also on the rate of heat loss from the
system. Unfortunately it is therefore not possible to determine and assign a stability
temperature to a substance as one can with melting points or flash points. A material
which is stable at some temperature in one situation may runaway from the same temperature
if the system, in particular the rate of heat loss, changes.

There are two extreme cases which can be considered in describing heat loss from a system
(Fig. 1). In the first, originally discussed by Semenov (ref. 1), the temperature is
assumed to be uniform throughout the reactant mass. This situation occurs in gaseous and
well stirred liquid systems where the rate of heat loss is governed by heat transfer at
the boundary.

In the second case, considered by Frank-Kamenetskii (ref. 2), the temperature distribution
is non-uniform and heat loss is controlled by heat transfer through the bulk. This occurs
in large unstirred liquid masses, powders and solids.

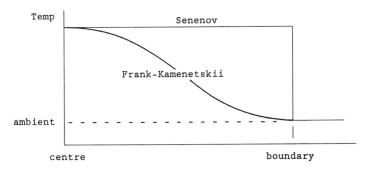

Fig. 1. Semenov and Frank-Kamenetskii Temperature Profiles

Systems with uniform temperature distribution
Semenov assumed a pseudo zero order exothermic reaction following an Arrhenius type rate
law, that is the rate of reaction and therefore the rate of heat production increases
exponentially with temperature. Thus for an irreversible nth order reaction A --> R at
constant volume V the rate of heat production Qr is given by

$$Q_r = V \, (- dH_r) \, k_o \, C_{Ai} \, \exp \, (-E/RT)$$

where dH_r is the heat of reaction, C_{Ai} is the initial concentration which is assumed to
remain constant for a limited time and k_o is the initial rate constant for the reaction
with activation energy E.

The rate of heat loss Qc is assumed to be governed by Newtonian cooling, that is it is
linearly dependant on the temperature difference, the heat transfer coefficient, U, and
area, A:

$$Q_c = UA \, (T - Ta)$$

Three cases for difference ambient coolant temperatures can be discussed (Fig. 2).

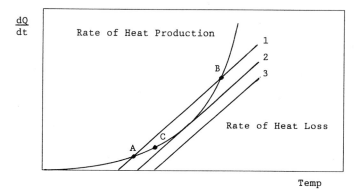

Fig. 2. Heat balance for Semenov type systems.

In the first case the rate of heat loss (line 1) intersects the exponential heat
production curve at two points A and B where the chemical heat production rate is balanced
by the heat removal capacity. The low temperature point A represents a stable situation
which can be illustrated by considering an increase in temperature to point C. At this
temperature the rate of heat loss is greater than the rate of heat production and the
temperature will return to point A. In contrast point B is unstable as any slight
increase in temperature will cause an increase in the rate of heat production not matched
by the rate of heat loss and an accelerating runaway will occur.

Line 3 represents the situation where the rate of heat loss from the system is always less
than the rate of heat production and a runaway reaction will always occur.

Line 2 describes the critical situation where the heat production is just equal to the
heat removal.

In addition since the rate of heat loss is dependant on the heat transfer coefficient and
area, a decrease in either will lead to a decrease in the slope of the line and a
reduction in the rate of heat loss from the system.

Systems with non-uniform temperature distribution
These systems cannot be described by a single temperature, rather the temperature is a
point function depending on the three spatial coordinates. The problem of analysing heat
transfer and hence of deriving the critical conditions for thermal explosions therefore
becomes extremely complex.

In essence the onset of runaway reactions in such systems depends not only on their size but also on their shape or symmetry. An increase in sample size reduces the heat loss from the reaction zone at the centre of the sample by acting as additional insulation. Since heat loss from the system occurs at the surface this change is proportional to the surface area whilst the heat generation is proportional to the mass or volume of the substance. Thus the ignition temperature decreases as the size increases.

Most workers base their analysis on the treatment by Frank-Kamenetskii with the basic equation:

$$\ln \left(\frac{\delta cr \, Ti^2}{r^2} \right) = M - \frac{N}{Ti}$$

Where δcr is the a dimensionless critical parameter that depends on the size and shape and boundary conditions of the material, Ti is the minimum ignition temperature for the particular system, r is the radius and M and N are constants that characterise the material such that

$$M = \ln \left(\frac{p(-dHr)EZ}{L R} \right) \quad \text{and} \quad N = \frac{E}{R}$$

where Z is the Arrhenius preexponential factor, p is the density and L the thermal conductivity.

This shows that a plot of $\ln (\delta cr \, Ti^2/r^2)$ against $1/Ti$ for a given substance in a series of containers of similar shape but different sizes, results in a straight line with slope of $- E/R$.

Further details can be found in the reviews by Gray and Lee (ref. 3) and Bowes (ref. 4).

A simpler treatment is given by Leuschke (ref. 5) which results in a relationship between the ratio of the volume, V to the surface area, SA and the ignition temperature such that:

$$\ln \left(\frac{V}{SA} \right) \propto \frac{1}{Ti}$$

This allows a quick and approximate correlation to be made for samples of different shapes and sizes.

FACTORS AFFECTING THERMAL STABILITY

The influence of heat transfer on thermal stability and the marked affect it can have on whether a runaway will occur is demonstrated by the following incident:

Following distillation at $150^{\circ}C$ the heating on a reactor containing distillation residues was switched off and the reactor was left to cool naturally before discharge. The temperature fell to $80^{\circ}C$ however 13 hours later, rapid decomposition occurred which took the temperature to above $200^{\circ}C$. Investigation showed that the still residues slowly decomposed at the temperatures involved and that they started to solidify at temperatures below $100^{\circ}C$. While the residues were molten the heat that was generated due to their decomposition was lost from the system, however as the residues solidified the heat transfer changed and the rate of heat loss at $80^{\circ}C$ was less than the the small rate of heat generation at this temperature, hence a runaway decomposition occurred.

In contrast to liquid reaction masses which contain all the components involved in their decomposition, air is often an essential ingredient in the exothermic activity of powders. Factors such as packing density and the distribution of interparticle voids that are not a function of the chemical constitution of the powder often control the initiation and rate of decomposition.

The majority of substances handled in the chemical industry are complex molecules or mixtures of substances and decompositions can involve consecutive and or parallel reactions. The rate of heat production depends on their chemical constitution and reactivity. Small changes in formulation can often markedly affect the decomposition process. Autocatalytics in which the first stage of a reaction produces a catalytic component in sufficient concentration to activate the second may also occur.

In addition contaminants can have a marked effect on thermal behaviour: An explosion occurred during spray drying of a material for which the lowest temperature at which ignition of 10mm layers of material (the thickest that would occur during operation) was measured as 350°C, well above the maximum temperature in the dryer of 220°C. Detailed examination of the material in the dryer showed contamination by the lubricating oil from the inlet atomiser. This contamination reduced the ignition temperature to 215°C and initiated the explosion.

Thus it can be seen that many factors affect the stability of substances. The material tested must therefore be representative of the substance handled in full scale manufacture and the tests used should simulate the practical conditions including the temperatures and time cycles that will occur.

DETERMINATION AND EVALUATION OF THERMAL STABILITY

The balance between the considerations discussed above concerning the rate of heat production and loss in a system are also applicable to the methods used to measure thermal stability and to the evaluation of the data obtained.

The heat generated in an exotherming mass is used in 3 ways as shown in Fig. 3.

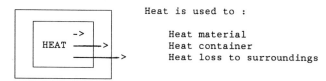

Fig. 3.   Schematic Representation of Heat Balance

Thus heat is used to raise the temperature of the mass, the temperature increase being dependent on the specific heat:

$$dT = \frac{dH}{Cp}$$

Heat is also used to raise the temperature of the sample container. This proportion of heat which is used in heating the container is called the Phi factor (6) which is expressed as:

$$PHI = \frac{\text{Heat Capacity Sample and Container}}{\text{Heat Capacity Sample}}$$

Finally heat is lost from the system to its surroundings.

Of these both the heat used to raise the temperature of the container and the heat lost to the surroundings vary between experimental test systems and the full scale situation. Thus increases in Phi will abate both the temperature rise and the rate of temperature rise observed in the test compared to the plant situation, similarly, if there is a higher rate of heat loss from the test system, exothermicity which is detected by measuring a temperature increase will not be observed in the test but will occur on the full scale.

This effect is shown in Table I which compares the rate of heat loss from different sized vessels with that in small scale apparatus.

The table also gives the half life which is the time taken for the temperature to fall to half its original value and also the time required for a 1°C temperature drop. It can be seen that the heat loss from the typical small scale experimental apparatus and screening tests used is far greater than occurs in plant items, the data obtained has therefore to be extrapolated. Tests using simple glass Dewars can simulate small plant reactors, up to 12.7m^3 (ref. 7), however to obtain data under conditions that represent larger reactors it is necessary to use adiabatic Dewar calorimetry (ref. 8).

Table I Heat loss from Vessels and small scale apparatus

	Heat loss W/Kg/K	t1/2 hr	Time for 1^{0}C loss at 80^{0}C (ambient = 20^{0}C)
2.5m^3 Reactor	0.954	14.7	21 min
5m^3 Reactor	0.027	30.1	43 min
12.7m^3 Reactor	0.020	40.8	59 min
25m^3 Reactor	0.005	161.2	233 min
10ml Test tube	5.91	0.117	11 sec
100ml glass beaker	3.68	0.188	17 sec
DSC/DTA	0.5-5		
10g Screening Tests	3-8		
1l glass Dewar flask	0.018	43.3	62 min
1l S.S. Dewar in Adiabatic Oven dT = -1 K	0.195	-	247 min

Explosivity screening

An essential first stage before both carrying out any testing and indeed before manufacture or handling any material, is to identify and exclude any substances, reaction masses, or residues that could have detonation or deflagration explosive potential.

A substance should be tested for detonating or deflagrating explosive properties if it either

a) contains chemical groups such as nitro, nitroso, azo, peroxy, acetylene which are known to confer explosive properties to a material.(see Bretherick (ref. 9) for a more complete list)
or
b) its oxygen balance which is a measure of the oxygen available in a molecule is more positive than -200. The oxygen balance is calculated by assuming complete reaction to carbon dioxide and water:

$$C_xH_yO_z + (x + \frac{y}{4} - \frac{z}{2})\ O_2 \quad \text{-->} \quad xCO_2 + \frac{y}{2}\ H_2O$$

$$\text{Oxygen Balance} = \frac{-1600\ (2x + y/2 - z)}{\text{Molecular Weight}}$$

Typical values for materials of known instability are nitrobenzene (-162.6), glyceryl trinitrate (+3.5) and dinitro toluene (-114.2).

Laboratory tests

All laboratory tests to evaluate thermal stability involve exposing a small quantity of the substance to elevated temperatures and monitoring any deviation in the sample temperature from the applied temperature.

Three temperature-time conditions are used (Fig. 4). **Programmed** in which the temperature is increased at a predetermined rate, _isothermal_ in which the sample is held for an extended time at an elevated temperature and _adiabatic_ in which the heat loss from the sample is minimised by the temperature of the oven following that of the sample. In all cases the temperature of the sample is monitored, a positive deviation from the applied temperature indicating the onset of exothermicity.

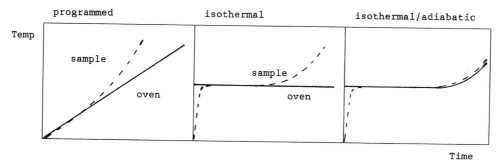

Fig. 4 Schematic diagram of temperature-time modes used in testing

DTA/DSC tests which used a small (5-50mg) sample are often used to give a preliminary indication of thermal stability of a material and if the mechanism of decompositions are known can be used to provide an estimate of the kinetics of the reaction (ref. 10 & 11). Gram scale screening tests have been developed by several workers (ref. 12 & 13) and several commercial versions are available, e.g. ARC, RADEX. An understanding of the sensitivity of the test used is essential in interpreting the data obtained and applying it to the full scale plant situation.

Dewar Calorimetry, one of the oldest and simplest calorimetric methods, in which the sample is held in a Dewar flask contained in an oven to further minimise the heat loss, continues to be widely used in the investigation thermal stability (ref. 7, 8 & 14). This is because the test apparatus can easily be set up to provide a thermal situation which closely simulates that which occurs in large scale reactors and storages. In addition the Dewar flask can be configured to mimic reactors by the provision of agitators and facilities for addition of reagents. Table I shows the sensitivity that can be obtained with such apparatus. The similarity to that which occurs in practice means that little if any extrapolation of the date is required.

Thermal stability of powders
The investigation of the thermal stability of powders is more complex because heat transfer is governed by thermal conduction through the bulk of the sample and the exothermic behaviour is often influenced by air availability. The tests described above can be used to give a preliminary indication of thermal stability, however as air availability is severely limited in these tests it is more common to use specific screening tests for powders which simulate the bulk, fluidised and layer situations which occur in practice (ref. 15 & 16). More accurate predictions of the behaviour of bulk quantities of powders can be obtained by determining the isothermal ignition temperature using wire baskets of differing sizes and fitting the data to the Frank-Kamenetskii theories described above (ref. 4 & 17).

CONCLUSIONS

Information on the hazardous properties of substances is a prerequisite to achieving safety in chemical production. Such data are not physical chemical constants and the values obtained vary markedly with the environment and particular plant situations. Thus there is no unique ignition/decomposition temperature. Test methods are available for investigating thermal stability, however, the complex reaction chemistry which occurs in such situations and the variations which can occur in heat transfer require an understanding of the plant conditions before the data can be evaluated.

REFERENCES

1. N.N. Semenov, <u>Z. Phys. Chem.</u> <u>48</u>, 571 (1928).
2. D.A. Frank-Kamenetskii, <u>Diffusion and Heat Transfer in Chemical Kinetics</u>, 2nd ed. Plenum Press New York/London (1969).
3. P. Gray, P.R. Lee, <u>Thermal Explosion Theory</u>, Oxidation and Combustion Reviews, ed D.F.H. Tipper, <u>2</u>, 1-183 (1967).
4. P.C. Bowes, <u>Self Heating: Evaluating and Controlling the Hazards</u>, H.M.S.O., London (1984).
5. G. Leuschke, <u>Self Ignition of Powdered Materials</u>, VF1B 5th International Fire Protection Seminar (1976) 1
6. D.I. Townsend and J.C. Tou, <u>Thermochimica Acta</u>, <u>37</u>, 1-30 (1980).
7. R.L. Rogers, <u>I.Chem.E. Symposium Series No. 115</u>, 97-107 (1989).
8. T.K. Wright and R.L.Rogers, <u>I.Chem.E. Symposium Series No. 97</u>, 121-132 (1986).
9. L. Bretherick, <u>Handbook of Reactive Chemical Hazards</u>, 3rd ed., Butterworths, London (1985).
10. R. Gygax, M.W. Meyer and F.Brogli, <u>Thermal Analysis</u>, <u>1</u>, 541-8 (1980).
11. G. Hentze, <u>Thermochemica Acta</u>, <u>72</u>, 127-138 (1984).
12. <u>Guidelines for Chemical Reaction Hazard Evaluation</u>, APBI, London (1989)
13. N. Gibson, R.L. Rogers and T.K.Wright, <u>I.Chem.E. Symposium Series No. 102</u>, 61-83 (1987).
14. R.L. Rogers, <u>Plant/Operations Prog.</u>, <u>8</u>, 109-112 (1989).
15. N. Gibson, D.J. Harper and R.L. Rogers, <u>Plant/Operations Prog.</u>, <u>4</u>, 181-189 (1985).
16. <u>Prevention of Fires and Explosions in Dryers</u> 2nd ed. I.Chem.E., Rugby (1990).
17. P.F. Beever, <u>I.Chem.E Symposium Series No. 69</u>, 4/X:1 (1981).

Fact-finding and basic data
Part II: Desired chemical reactions

Ruedi Gygax

Ciba-Geigy AG, CH-4002, Basel, Switzerland

Abstract - A major hazard in running chemical processes is loss of cooling. Its immediate consequence is a rapid temperature increase, corresponding to the potential energy represented by whatever amount of unreacted starting materials is present at the time of the loss of cooling. Depending on the temperature level reached, the runaway of the desired reaction may have further consequences. In particular, it can be followed by the runaway of an often more exothermic decomposition reaction. To rule out these scenarios, they must be studied. Basic data obtained by thermal analytical experiments serve to characterize runaway scenarios and allow to design processes which are tolerant towards deviations from operating conditions.

INTRODUCTION

Important data of Hazard Analysis concern the knowledge of the two components of risk, severity and probability. As has been shown in the first part, the energy potential inherent in the materials processed is the dominant factor which determines severities. Probabilities are linked with mechanisms of activation of these energy potentials. In order to judge upon such mechanisms, they must be identified and studied (ref. 1).

PHYSICAL OPERATIONS VERSUS CHEMICAL REACTIONS

If storage or **physical operations** are the subject of Hazard Analysis the major mechanism of activation is an improper balance between the heat produced by undesired chemical reactions and the heat dissipation rate, which may be very low. The hazard stemming from the unwanted chemical reactivity of the materials processed can be coped with by the following procedure (ref. 2):

(1) identification of the conditions under which undesired chemical transformations do produce dangerous rates of energy releases and

(2) definition of safe operational limits sufficiently away from such conditions.

Contrastively, in **chemical syntheses** the very goal is chemical transformation and the energy release accompanying it is coped with by design in the operational plant. It is important to note, therefore, that the necessary approach of assessing the thermal hazards of chemical processes is fundamentally different from the procedure described above for physical operations.

While also for chemical synthesis the severity of a risk often still stems from high heats of undesired chemical reactions, the modes of activation of the decomposition energies are typically linked with the loss of control of the process employed to run the desired reaction (ref. 3).

BASIC CONCEPTS ABOUT THE SAFETY OF CHEMICAL PROCESSES

For assuring a synthetic process to be safe, two basic conditions must be fulfilled:

Postulate (1)   The heat release during the process must be known and the process equipment must be designed such that dissipation of the released heat can be assured at all times.

Postulate (2)   **The process must be sufficiently tolerant towards deviations from the operating conditions. E.g. a complete breakdown of the cooling capacity, which is taken as prototype of deviation, must not lead to intolerable consequences.**

Accordingly, "basic safety data" must provide a link to the assessment of how well the two postulates are coped with by a given process. This is done by keeping track of the energy potentials present in the reactors and by studying the runaway profiles to be expected in the case the energy potentials are uncontrollably released.

Taking such a data-oriented approach to characterizing runaway scenarios does not necessarily mean that plentiful and very accurate data about them are required. It is more meaningful, to know by the order of magnitude, that conditions of a dangerous runaway scenario are far away from actual conditions than to determine very accurate data about a close to critical situation. The goal must be to identify reaction processes, which bear a high thermal risk and then to modify them into processes with little inherent accumulation of energy potentials.

Considerations about the safety of synthetic chemical processes are very relevant at the stage of process design and, vice versa, process optimization must include safety aspects. Integrating hazard assessment procedures and process design is essential. A means of achieving this is to refer to "basic data", which give information on the thermal behavior of the reacting system taking into account the process conditions investigated.

Next, we shall outline the key points characterizing the runaway behavior of the system investigated and then expand the discussion to experiments and their interpretation with respect to the key points.

OUTLINE OF AN ASSESSMENT PROCEDURE OF THERMAL HAZARDS.

To satisfy the first postulate of controlling the running plant, the fact finding procedure involves to provide data on the energy release determined under the true operating conditions of the process, and then to deal with the heat dissipation properties of the production equipment, taking into account the laws of scale-up. Required basic data include:

the **heat evolution of the reaction** as a function of the process time, and

the **heat dissipation behavior of the reactor** under the respective process conditions.

These requirements are obvious and are dealt with by genuine engineering knowledge.

In order to address the second postulate, an educated assessment of deviation paths is called for. Thermal data on both the desired reaction and the undesired decompositions enable the investigator to sketch worst case runaway scenarios, which must be expected under given failure modes (e.g. breakdown of cooling). They serve as a reference to the judgement about the risk associated with them.

In order to characterize a failure mode scenario, data on the main features of runaway profiles are determined. They comprise (see fig. 1):

Feature (1)   The maximal temperature which will be reached after the cooling breakdown if, in a first consideration, only the heat of the desired reaction is taken into account.

Feature (2)   An estimate of the end temperature of the runaway of secondary, undesired reactions.

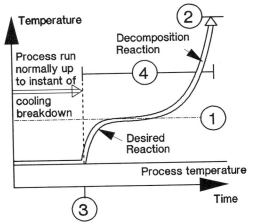

Fig. 1. Schematic runaway profile

Feature (3)   The instant within the process, at which cooling breakdown leads to the most consequential runaway profile.

Feature (4)   The time frame, within which the projected runaway reaction is predicted to occur.

EXEMPLARY SAFETY ANALYSIS OF DIFFERENT TYPES OF CHEMICAL PROCESSES.

In the following, a simple reaction will be considered as an example. It consists of two reactants, used in equimolar amounts, which form a product. The product undergoes a fairly exothermic decomposition reaction at elevated temperatures. Both experiments elucidating the basic thermal data of the reaction system and design considerations for safe processes will be discussed. All the data have been compiled on the basis of a simple kinetic model to provide consistency throughout the discussion.

The static picture (Severity)
When looking at a reaction system, it is reasonable to first consider the overall energy potentials involved. They are easily measured and, if they are small, the assessment procedure can be abbreviated. A priory information about the energy potentials stems from known heats of reaction specific to the reaction type and from knowledge about contributions to the decomposition potentials by the individual groups present in the molecules.

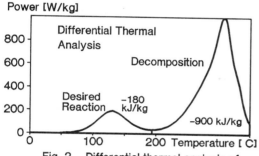

Fig. 2. Differential thermal analysis of example reaction (Scan = 4°C/min).

Energy potentials can be measured, e.g. by reaction calorimetry (see below) and by differential thermal analysis DTA[1]. Figure 2 displays the DTA result of a mixture of starting materials of our model reaction. Two peaks are seen. Typically, as the temperature is gradually increased by the instrument, the peak which corresponds to the desired reaction occurs first. The larger peak at higher temperatures indicates a strong decomposition of the product mixture being formed by the reaction. It is important to note, that the exact location of the peaks on the temperature axis and their shape depends on experimental parameters, and – although it does bear some information also on reaction dynamics – the huge difference in scale and operation between experiment and plant conditions must be carefully taken into account when attempting to assess the reactivity of the system investigated directly from scanning DTA. This will become more evident, when looking back at the example DTA-curve later in the discussion.

However, the DTA result gives a direct information about the energy potentials involved. The measured overall energy potential is a quite direct measure of the severities implicated. This is most evident if the energy potentials are converted to temperature rise potentials by dividing them by the heat capacities[2]. The obtained values can be used to estimate secondary consequences, such as evaporation rates, pressure increases, etc., depending on the specific properties of the materials and equipments used. In the model case, the temperature rise potentials, and with them, the severities are high[3]. The safety of the reaction process entirely depends on whether process conditions can be found, which exclude mechanisms for triggering the total release of all the energy potentials present.

---

[1] Adequate sample preparation, pressure tight sample containers and a sufficiently small scale (milligrams) are essential for quantitative results; see ref. 4.

[2] Heat capacities of the mixtures can be estimated with sufficient accuracy from tabulated values of the used or of structurally related compounds. If necessary, they can be measured by reaction calorimetry and by DTA, etc.

[3] With a heat capacity of 1800 $J\ kg^{-1}\ K^{-1}$, temperature rise potentials of 100°C and 500°C are calculated for the desired and follow-up decomposition reactions, respectively.

The dynamic picture (Probability)

In order to avoid such mechanisms, their dynamics must be studied. This, essentially, involves two questions:

(T_1) How high can the temperature rise without triggering significant release of the decomposition potential?

(T_2) How high will the temperature rise, if - in a given process - all the energy potential of the desired reaction present at any given time is converted into temperature increase?

If temperature (T_1) is low compared to (T_2), a runaway of the desired reaction is predicted to trigger the release of the high decomposition potential on a relevant time scale, an unacceptable situation.

If temperature (T_1) is high relative to (T_2) and if no secondary effects (i.e. evaporations, etc.) must be feared, the consequences of loss of control will be limited, and the process bears much less inherent risk, even though the same high decomposition potential as in the former case may be present.

Runaway of the decomposition reaction (Question about T_1). The logic of answering the first question is the same discussed earlier about risk assessment of physical operations. Essentially, the fact finding procedure calls for experiments which result in the prediction of runaway times under completely uncooled (adiabatic) conditions as a function of initial temperatures[4]. To determine which runaway times are acceptable, again, specific information on plant equipment and management of emergencies are important and, judgement beyond mere fact finding is involved.

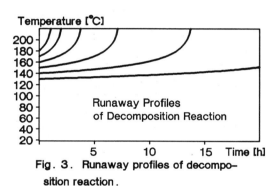

Fig. 3. Runaway profiles of decomposition reaction.

Figure 3 reports adiabatic runaway profiles of the example decomposition as a function of temperature. E.g., for a typical plant situation a reasonable argument would hold that, concerning the hazard of decomposition, temperatures up to around 130 °C are acceptable in the example case, as it is highly unlikely, that a cooling failure could go undetected and that adiabatic conditions could prevail for the relatively long expected runaway time of more than 20 hours from around 130 °C. Oppositely, if the temperature were to rise to, say, 180 °C by runaway of the desired reaction (or by accidental heating), this event would trigger the release of the high decomposition potential within a time of about one hour and consequently without much chance of breaking the prevailing adiabatic conditions.

The heat evolution accompanying the desired process. A discussion of the thermal reaction dynamics of the desired reaction starts with the expected heat evolution. Some prior information about the heat evolution may be derived from the overall heat of reaction and knowledge of the rate of conversion of the reaction.

The reaction calorimeter is an instrument, which has been specifically designed to provide data about heat evolutions, while the process is being run according to the authentic plant procedure (ref. 6). E.g., consider a batch reaction process, i.e. a process in which all the starting materials are loaded into the reactor at the start. The result of a reaction calorimetry experiment about a batch process of the example reaction run at 80 °C is given in fig. 4. From these type of data, it is obvious how much heat must be dissipated in the plant, if no temperature rise is desired (Postulate No. 1). At the beginning, the batch process of fig. 4 exhibits a very high heat release rate, which - in this case - may be too difficult to cope with even in an operational plant.

Summing up all the heat measured in the reaction calorimeter over the process

---

[4] The use of isothermal DTA experiments in connection with a calculation of "Times to Maximum Rate" may be used for these predictive estimates. For a discussion see the preceding contribution by Dr. Rogers and refs. 2,5.

time allows to determine the integral heats of reaction[5].

Temperature rise potential by the desired reaction as a function of process designs (Question about T_2). Next, the question of the runaway potential of the desired process as a function of the process time is addressed.

Batch Process: In the batch process of fig. 4, all the reagents are present in the beginning. As time progresses, conversion of the reactants takes place while the corresponding heat of reaction is released and measured in the reaction calorimeter (fig 4a.). We can therefore take the sum of the heat set free up to a certain point to measure the degree of conversion. Similarly, the potential remaining in the reaction vessel at any considered time, can be obtained by integrating the heat released in the reaction calorimetry experiment from this time to the end (shaded area in fig. 4b). Thus, without interrupting the experiment at any time, we can deduce the potential reaction energies of the reactants present in the reactor at any time.

If, again, we divide the values of these energy potentials by the heat capacities, we obtain the temperature rise potential of the desired reaction as a function of the process time.

Starting from the process temperature of 80°C and adding the temperature rise potential of the desired reaction, results in the end temperature of the expected runaway (fig. 4c). If the system turns adiabatic at the very start of the process, 180°C will be reached by runaway of the desired reaction alone. Only after about two hours, the potential runaway temperature of the desired reaction falls below 130°C, the range termed critical in a previous argument.

Figure 4d simultaneously displays the final runaway temperature of the desired reaction and the runaway decomposition profiles to show their relative importance on the temperature scale. Clearly, at the beginning, a runaway of the desired reaction is critical. It can only be avoided by sufficient cooling. Thus the batch process, in the considered example case, is not tolerant to loss of cooling.

Ramped Batch Process: The example batch process obviously lacks sufficient safety in its initial phase. One attempt to improve it is to use a lower initial temperature and ramp the temperature to accelerate the reaction in its later phase. The reaction calorimetry analysis of such a process is displayed in fig. 5. Again, the directly measured heat evolution curves (fig. 5a) are used to derive the energy potentials and corresponding temperature rise potentials for all times. The considered runaway now starts at a lower temperature,

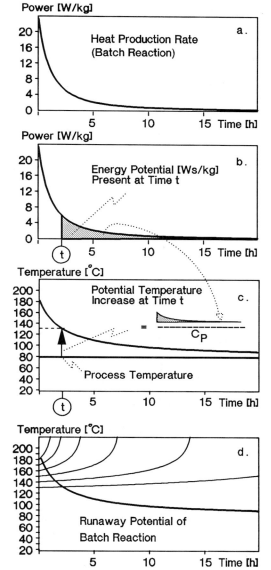

Fig 4. Batch reaction process; derivation of potential runaway temperatures (c, d) from heat production rate curve measured by reaction calorimetry (a, b).

---

[5] Integral heats measured may also contain heats of physical transformations, e.g. heats of mixing, latent heats etc. Procedures do exist to separate individual contributions, if necessary.

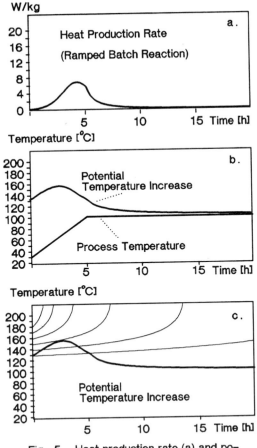

Fig. 5. Heat production rate (a) and potential temperature increase (b, c) of a batch reaction with a temperature ramp.

but also, the initial conversion rate has been decreased. As shown in fig. 5b, the most critical time for the cooling interruption to occur has been shifted from the start to some time later on. The end temperature to be expected in the case the desired reaction does runaway adiabatically at this most critical moment, is now lower than in the batch reaction of fig. 4, but still reaches into the temperature region in which the decomposition reaction is critically active (fig. 5c).

By introducing temperature ramps with varying set points and slopes, many degrees of freedom are available for optimizing the process further. The methodology discussed above allows the highest end temperature of an adiabatic runaway of the desired reaction to enter the optimization procedure as the relevant thermal safety criterion. In a ramped batch process, the heating absorbs part of the heat of reaction and thus, during the ramp, the state of no energy interchange represents some "cooling" relative to the temperature aimed at by the process prescription.

Semi-Batch Processes: Another way to reduce the potential present at the beginning of a process is to feed one of the reactants into the reactor only gradually. In the ideal case, a feed controlled process can be obtained, in which the reactant is consumed by the reaction at exactly the rate with which it is fed to the reacting system. No runaway potential of the desired reaction whatsoever is accumulated in this ideal case (titration). The control of the temperature remains entirely with the operator, as he can stop the heat production by stopping the feed if this is required to maintain the set temperature.

Unless the reaction rate is very fast, in reality, some reactant accumulation always occurs. Figures 6 and 7 display the reaction calorimetry investigations of two semi-batch processes chosen among a large variety which can be imagined. The thermal potential present in the reactors are calculated by keeping track of the input of potential - running from zero at the start of the feed to the full potential at the moment when the total of the fed reactant is equimolar to the reactant initially present in the reactor and deducing the sum of the heat output measured up to the considered time[6].

For both of the semi-batch processes discussed, any adiabatic runaway of the desired reaction will be limited to temperatures judged uncritical, if the feed is interrupted as the systems become adiabatic. Other than the batch processes, the example semi-batch processes have a sufficiently large built-in gap which separates a runaway of the desired reaction from the triggering of the decomposition reaction.

As can be seen by comparing the two cases, there is a trade-off between a higher process temperature and a lower potential temperature runaway. Interestingly, of the two semi-batch processes the one run at higher temperature has a slightly higher built in safety gap. This again illustrates that, other

[6] In the example processes, the point of equimolar feed coincides with the end of the feed. When deriving the potential runaway temperature, the varying mass must be taken into account for obtaining an accurate value. Often, however, the neglect of this complicating factor does not introduce too large an error.

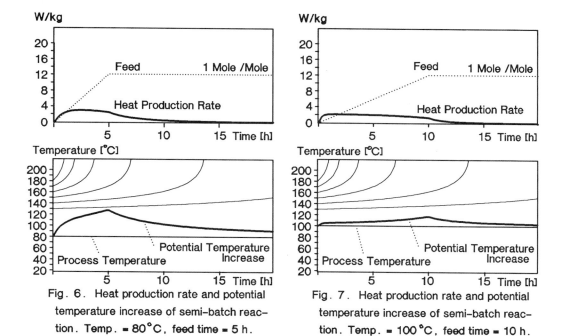

Fig. 6. Heat production rate and potential temperature increase of semi-batch reaction. Temp. = 80°C, feed time = 5 h.

Fig. 7. Heat production rate and potential temperature increase of semi-batch reaction. Temp. = 100°C, feed time = 10 h.

than in the case of physical operations, a lower temperature for a desired chemical process is not always the safer one (ref. 7).

CONCLUSION

Our discussion of "basic data" of desired reactions has necessarily been exemplary. This implicates, that it is not reasonable, to collect many isolated data in a too routine manner. Beyond employing the general methodology to study runaway mechanisms built into given processes, testing and assessment procedures should be adapted to the needs of individual processes. The assembly of "basic data" is an iterative process, in which at each state information available from previously measured data determines which question should be investigated next.

Taking an approach integrating process safety considerations with process design leads to the benefit of obtaining better processes. Moreover, once the process design has been fixed, referring to the runaway behavior of the systems studied allows to assess the relative risks inherent in the given process. Thus it allows to adequately design necessary measures for controlling the reaction systems with particular reference to failure mode conditions of the plant and equipment determined elsewhere in the risk analysis procedure.

REFERENCES

1. Thermische Prozess-Sicherheit, ESCIS-Series 8 (1988); available from SUVA, CH-6001 Lucerne; English translation to be published.

2. R. Gygax, Chem. Eng. Progress 53 (1990)

3. H. Fierz, P. Finck, G. Giger, R. Gygax, The Chemical Engineer 400 9 (1984); R. Gygax; Chem. Eng. Sci. 43 1759 (1988).

4. F. Brogli, R. Gygax, M. Meyer, Proceedings of the 6th Int. Conf. on Thermal Analysis, Bayreuth, Birkhäuser, Basel, 541, 549 (1980).

5. D.I. Townsend, J.C. Tou, Thermochimica Acta 37, 1 (1980).

6. W. Regenass, W. Gautschi, H. Martin, M. Brenner Proc. ICTA 3 3 (1974); T. Hoppe, B. Grob, Intl. Symp. Runaway Reactions AICHE 132 (1989).

7. P. Hugo, J. Steinbach, F. Stoessel, Chem. Eng. Sci. 43 2147 (1988).

Fact-finding and basic data
Part III: Fire and explosion

Helmut Schacke

Bayer AG, Leverkusen, Germany

INTRODUCTION

The hazard potential - its origin from the reactivity, in this case the undesired and unloved reactivity - of the chemical substances and the necessity to get down with this problem have already been introduced and emphasized by the preceeding papers (Fact Finding and Basic Data, Part I and Part II).

In terms of flammability and explosivity the **hazard potential** of chemical processes and plants is given by the ability of many chemical substances to undergo - as the fuel - **exothermic oxidation reactions** in the gas phase which represents a certain **dispersed state**. These reactions are accompanied by considerable energy release noticable as heat, radiation, or pressure-build-up (and in some cases also by the formation and release of dangerous combustion products). In contrast to the slower stationary combustion (burning), for an explosion an independent fast propagation of the reaction zone (flame) through the free volume with the reactive mixture is essential. This latent hazard potential may be triggered off by an initiator, the ignition source (Fig.1). So the unholy trinity of **"fuel"**, **"oxidator"**, and **"ignition source"** fairly well describes the scope for which basic data are needed.

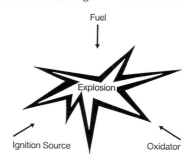

Fig. 1. Explosion Factors

COMBUSTION PROPERTIES/CONDITIONS

Since it is not the chemical substance as the fuel by itself which forms the hazard potential but its proper mixing with the oxidator (usually the oxygen of the air) in a **first step** the characteristics of the system fuel/oxidator have to be defined. These should answer the question whether the substances may give rise for fire or explosions. For the assessment of the combustion behaviour of chemical substances important data are e.g.:

1. the **Burning Velocity** of (coherent) solids or powder accumulations, or - in an approximation - the so called **Burning Index BZ**, which can be measured by a very plain method which is shown in Fig.2. This method consists of a simple fire train test where one can assign a burning index between BZ 1 (e.g. sodium chloride) and BZ 6 (e.g. black powder) just by visual assessment of what happens to the heap of powder after ignition by a small flame (ref.1),

2. the **Flashpoint** T_f of a liqiuid as the lowest temperature at which the liquid will flash when ignited and will form explosive vapour/air mixtures, measured again with a relatively simple method, shown in Fig. 3 which also gives some typical data (ref. 2,3),

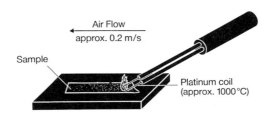

BZ 1 Sodium Chloride .... BZ 6 Black Powder

Fig. 2. Burning Index/Test Apparatus

3. the **Explosion Limits**, **Lower Explosion Limit LEL** und **Upper Explosion Limit UEL**, as concentrations of gases, vapours, or dusts dispersed in an oxidizing gas phase, including the limiting oxidator concentration LOC (since there are oxidator concentrations too low to enable flame propagation).

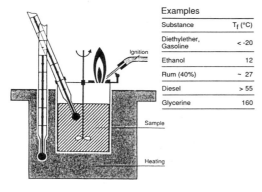

Examples	
Substance	T_f (°C)
Diethylether, Gasoline	< -20
Ethanol	12
Rum (40%)	~ 27
Diesel	> 55
Glycerine	160

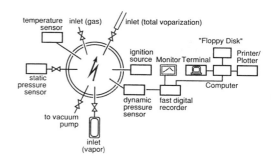

Fig. 3. Flashpoint/Test Apparatus

Fig. 4. Explosion Limits/Test Apparatus

The determination of those fuel and oxidator concentrations - which in fact are explosion conditions - already require more sophisticated experimental methods, what one may realize from Fig. 4 (ref. 4): a pressure vessel in which gas/vaour explosions are deliberately introduced under variation of the mixtures' composition; controlled, monitored, recorded, evaluated and documented by means of some electronic devices (similar method used for dust explosion conditions, ref. 1). Some data are given in Table 1 (ref. 3).

For some purposes more advanced data are needed, for instance explosion limits for oxygen enriched atmospheres, elevated pressures or temperatures, expediently put down in a ternary diagram as shown for the system $C_2H_4/N_2/O_2$ (see Fig. 5).

Chemical Substance	Lower Explosion Limit LEL		Upper Explosion Limit UEL		Limiting Oxygen Concentration LOC
	Vol. %	g/m³	Vol. %	g/m³	Vol. %
Hydrogen	4	3.3	75.6	64	5
Methane	5	33	15.0	100	12.1
Benzene	1.2	39	8.0	270	11.2
Propane	2.1	39	9.5	180	11.8
Polyethylene*)		30	kg/m³ – range, for safety considerations usually not of relevance		10
Wheat Flour*)		60			11
Aluminium*)		30			6

*) dependent on particle size distribution

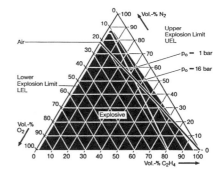

TABLE 1. Explosion Characteristics of some Chemicals (p_0 = 1 bar, $T_0 \sim$ 300 K)

Fig. 5. Explosion Limits $C_2H_4/N_2/O_2$ (p_0 = 1 bar/16 bar, $T_0 \sim$ 300 K)

IGNITION REQUIREMENTS

In a **second step** one will have to find out the (minimum) requirements to **activate** the fire and explosion hazard, i.e. the flammable or explosive system's characteristics for being ignited. Relevant to this task one will consider e.g.:

1. **Self Ignition** due to exothermic reactions, e.g. in powder accumulations. These are reactions which must be treated as an at least three parameter problem - temperature, volume, and (induction) time - a problem which has been dealt with in the previous papers (Fact Finding and Basic Data, Part I and Part II),
2. the **Auto Ignition Temperature AIT** as the lowest surface temperature to start ignition of an explosive mixture; this temperature being measured by an again simple method of feeding

a small amount of fuel into a heated Erlenmeyer glass vessel with temperature control and visually monitoring the ignition process (Fig. 6, ref. 5),

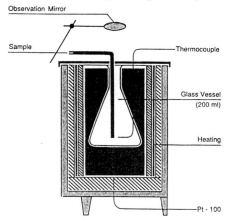

Fig. 6. Auto Ignition Temperature/
Test Apparatus

Chemical Substance	Auto-Ignition Temperature AIT (°C)	Minimum Ignition Energy MIE (mJ)	Maximum Experimental Safe Gap MESG (mm)
Hydrogen	560	0.019	0.29
Ethylene	425	0.07	0.65
Propane	470	0.25	0.92
Carbon disulfide	95	0.009	0.37
Wheat Flour	~450 (G.G.)	~100	
ABS	~450 (G.G.)	~10	
Sulphur	~250 (G.G.)	<1	

*) dependent on particle size distribution
G.G. = Godbert Greenwald Apparatus

TABLE 2. Ignition Requirements of
some Chemicals

3. the **Minimum Ignition Energy MIE** as the lowest (electrical) energy of an ignition source to start ignition of an explosive mixture. (This energy correlates strongly with the so called Maximum Experimental Safe Gap MESG which describes the limiting geometrical boundaries for the propagation of flames: Flames may be "quenched" when physical distances become too small. This will be referred to later on again.) The experimental set-ups are too complex to comment on here, so only some selected values of respective data will be given for comparison (Table 2, ref. 3).

EFFECTS (OF EXPLOSION)

In a **third step** we will focus on the explosive systems's behaviour after ignition thus indicating the physical explosion effects. These effects can be characterized by data as

1. the **Maximum Explosion Pressure** p_{max} (in an enclosed volume) and
2. the **Maximum Rate of Explosion Pressure Rise** $(dp/dt)_{max}$. A schematic illustration will give Fig. 7: After ignition of an explosive mixture in an enclosed volume there will be a steep pressure rise, followed by a shallower pressure decrease. The maximum pressure will be obtained for approximately stoichiometric mixtures. This pressure is of the order of 5 to 10 times the initial pressure and is thermodynamically controlled, whereas the rate is governed by the combustion kinetics. For experimental investigation the explosion chamber shown before (Fig. 4) would be the suitable apparatus. Here - as well as for stationary combustion (fire) - the

3. **Heat of Combustion** Δ **H** is the key term which for most fuels amounts to some ten kJoules per kilo gram (energy release manifests itself in pressure (rise) and radiation).

4. Also the **Maximum Experimental Safe Gap MESG**, a term already mentioned before, may be considered in view of explosion behaviour, although as a "negative" effect: It is a measure of the terminating condition to stop explosion (Table 2., ref. 6).

The kind of data given up to now exclusively describe characteristics of chemical substances. They can be obtained by lab-scale experimental investigations, in a few cases also by calculation methods. Some of the experimental methods have been **standardized**.

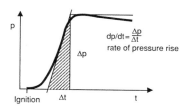

Chemical Substance	Maximum Explosion Pressure p_{max} (bar)	Maximum Rate of Pressure Rise (dp/dt) max (bar/s)
Hydrogen	7,1	550
Methane	7,4	55
Propane	7,4	75
Polyethylene*	9,0	200
Wheat Flour*	7,4	42
Aluminium*	11,5	500

* dependent on particle size distribution

Fig. 7. Explosion Characteristics of
Some Fuel/Air-Mixtures
(p_0 = 1 bar, $T_0 \sim$ 300 K, V = 1 m^3)

The type of experimental equipment ranges from rather simple (screening methods) to fairly sophisticated, with the experimental data always being subject to the drawback of **not** describing **natural constants**. Nevertheless, these data in many cases allow for a very useful **classification and rating** of chemical substances (Table 3.).

STANDARDIZED METHODS

allow

CLASSIFICATION

e.g.

Method/Data	Classification
T_f	AI, AII, AIII, B (in FRG)
AIT	Temperature Classes T 1 – T 6
MESG/MIE	Explosion Groups I, II A – II C
$(dp/dt)_{max}$	Explosion Classes K_{St}, K_G

Table 3. Classification

PROCESS AND PLANT PARAMETERS

As pointed out by the preceeding papers, it will yet not be sufficient for risk assessment having arrived at this base set of data. Still needed are **transformation rules** from lab-scale to "real life" and the knowledge about influencing factors and, furthermore, one will rather have to compare the substances' data with the characteristics of the chemical process and the plant, also and in particular for failure states.

Without going into details it can be stated that there are quantitative transformation rules which Table 4 gives a schematic view on. Here "+" indicates parallel effects, "-" inverse effects, and "0" indicates invariance. As an example: Increasing temperature makes the maximum explosion pressure decrease ("-"), or increasing initial pressure will lead to an extended upper explosion limit ("+").

Substance Safety Charact. (Lab-Data)	Real Plant Parameters			
	Temperature	Pressure	Dispersion	Volume
T_f		+		
LEL	–	–	–	0
UEL	+	+	(+)	0
LOC	–	–	–	0
AIT	–	–	–	–
MIE	–	–	–	0
P_{max}	–	+	0	0
$(dp/dt)_{max}$	+	+	+	–
MESG	–	–	(–)	(0)

"+" = parallel effect, "–" = inverse effect, "0" = invariant

TABLE 4. Transformation Rules

- Hot surfaces
- Flames, hot gases
- Mechanical sparks
- Electric installations
- Transient currents
- Lightning
- Static electricity
- Electromagnetic waves (high frequency)
- Electromagnetic waves ($3 \cdot 10^{11}$ Hz – $3 \cdot 10^{15}$ Hz)
- Ionizing radiation
- Ultra sonics
- Adiabatic compression, shock waves
- Chemical reactions

TABLE 5. Ignition Sources

The information about process and plant parameters not nessessarily in all cases has to consist of quantitative numbers. In a **fourth step** information must be available about e.g.:

1. the **presence and amount of oxidizable substances**
2. the **degree of dispersion** (for instance it makes a great difference whether one handles a solid log of wood or have it cut into saw dust; degree of dispersion may also stand for possible **flammable gas concentrations** or **vapour concentrations**, e.g. according to process temperatures exerted on flammable liquids), and last not least
3. the **generation** and characteristics of possible **ignition sources** (via electrical and electrostatic data, mechanical parameters, apparatus materials and equipment design).

This very important information eventually has to answer the question whether the process and the plant in normal operation or due to malfunction are prone to generate ignition sources which meet the ignition requirements of the flammable substances or explosive mixtures. There is a large variety of possible ignition sources, some of which being very obvious, others but only seem a little odd and academic (Table 5, ref. 7).

● Process, e.g.:
  – maximum temperatures
    (hot surfaces, self ignition)
  – conveying velocities (electrostatics)
  – product separation (electrostatics)

● Plant, e.g.:
  – apparatus (high temperatures
    induced by moving parts)
  – construction materials
    (mechanical sparks, electrostatics)
  – equipment
    (electrical sparks, electrostatics)

Fig. 8. Example: Ignition Source

A closer look to process and plant reveals the necessity for information e.g. about maximum temperatures leading to self ignition or about specifics of powderous product separation giving rise for incendiary electrostatic discharges (Fig. 8). In terms of electrostatics e.g. data about electrostatic conductivity, field strength, and resistance to ground will be necessary (ref. 8, 9). Construction materials combinations may be unsuitable because of possible incendiary frictional sparks. Electrical and PCI equipment may heat up or give electrical sparks of sufficient energy (ref. 10).

CONCLUSIONS

With this four-step-procedure (Fig. 9) a straight forward approach for "Fact Finding and Basic Data" is developed. The base set of data on chemical substances in terms of fire and explosion, combined with the information/data on process and plant parameters will provide a sound tool for the assessment of the fire and explosion hazard. At the same time they serve as a basis for the design of appropriate countermeasures which can be grouped into three areas. These measures aim at

Define Materials Data!

● STEP 1 Combustion
          Properties/Conditions

● STEP 2 Ignition Requirements

● STEP 3 Effects of Explosion

Define Plant and Process Parameters!
(STEP 4)

● Presence and Amount of Oxidizable
  Substances

● Degree of Dispersion
  (possible concentrations)

● Generation/Characteristics of
  Ignition Sources

Fig. 9. 4-Step-Procedure

1. the elimination of at least one of the explosion (fire) factors (Group 1 or 2), or
2. the minimization of their coincidence (by a probabilistic concept, ref.7, 11; Group 1 and 2) or, alternatively, at
3. the reduction of the dangerous consequences of a fire and an explosion down to a tolerable degree (Group 3). This last type of measures for explosion protection will generally reqire the implementation of (explosion) pressure proof apparatus and equipment.

In Fig. 10 this grouping is expressed, accompanied by a collection of important data and their application for the respective groups and countermeasures (Table 6).

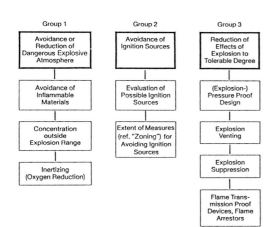

DATA/INFORMATION		APPLICATION		
Chemical Substance	Process/Plant	Group 1	Group 2	Group 3
Explosion Limits	Dispersion/Concentrations/Pressures/Ventilation/Dilution	●		
Limiting Oxygen Concentr.		●		
Flashpoint	Temperatures	●		
Auto Ignition Temperature	Temperatures Apparatus Geometry		●	
Minimum Ignition Energy	Conveying Velocity/ Construction Materials		●	
Max. Explosion Pressure	Apparatus Strength			●
Max. Expl. Press. Rise	Volumes/Venting Areas			●
Max. Experim. Safe Gap	Electrical Equipment/ Volumes/Pipe length		●	●

Fig.10. Basic Principles of
        Explosion Protection

TABLE 6. Application of Data/Information

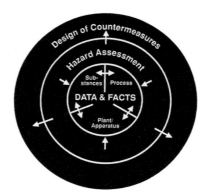

Fig. 11. Fact Finding and Basic Data

It is the **facts and data base** without which hazard identification and assessment, risk analysis and risk reduction, in general a safe design and a safe operation of a chemical plant will not be possible (Fig. 11). Thus "Fact Finding and Basic Data" form the starting point of all safety procedures and their integration into chemical production.

REFERENCES

1.   Test Methods for the Determination of the Safety Characteristics of Dusts, VDI-Guideline 2263, Part 1, VDI-Verlag, Düsseldorf (1990)

2.   e.g. DIN 51758, ASTM D 93, BS 2839

3.   data taken from relevant data collections such as
K. Nabert, G. Schön, Sicherheitstechnische Kennzahlen brennbarer Gase und Dämpfe, 2. Auflage, 5. Nachtrag, Deutscher Eichverlag, Braunschweig (1980);
BIA-Handbuch, Berufsgenossenschaftl. Institut für Arbeitssicherheit, Erich Schmidt Verlag, Bielefeld (1985);
and tables in ref. 7,8,x

4.   M. Müller, H. Schacke, C.D. Walther, Proceedings of the 6th International Symposium "Loss Prevention and Safety Promotion in the Process Industries", Vol. IV, p. 104-1 to 104-12 Norwegian Society of Chartered Engineers, European Federation of Chemical Engineering, Oslo (1989);
see also DIN 51649

5.   DIN 51794, BS 4056, ASTM E 659-78,

6.   IEC-Publication 79-1 A

7.   Richtlinien für die Vermeidung der Gefahren durch explosionsfähiger Atmosphäre mit Beispielsammlung, Explosionsschutz-Richtlinien, Richtlinien, No. 11, Berufs-genossenschaft der Chemischen Industrie, Heidelberg (1989)

8.   Richtlinien für die Vermeidung von Zündgefahren infolge elektrostatischer Aufladungen, Richtlinie No. 4, Berufsgenossenschaft der Chemischen Industrie, Heidelberg (1989)

9.   G. Lüttgens, M. Glor, Understanding and Controlling Static Electricity, expert verlag, Ehningen (1989)

10. VDE 0171, DIN 57165/VDE 0165, DIN EN 50014 to 50020, DIN EN 50039

11. H. Schacke, in Safety Standards and Regulations for Chemical Plants in Europe: A Comparison, Praxis der Sicherheitstechnik, Vol. 2, p. 151-162, DECHEMA, Frankfurt (1988)

x.   general, e.g.: H.H. Freytag, Ed., Handbuch der Raumexplosionen, Verlag Chemie, Weinheim (1965);
W. Bartknecht, Explosionen, Springer Verlag, New York (1980)

Identification of hazards for the purpose of safe chemical production

Dr. Volker Pilz

Bayer AG, Leverkusen, Germany

1. INTRODUCTION

We use chemical processes to turn raw materials into new products with interesting properties via chemical reaction and physical treatment. For such processes we have to use chemically reactive substances.

However, because the materials are reactive, they have hazardous properties such as

- toxicity,
- thermal instability,
- flammability and
- explosibility

which may cause a severe hazard to human beings and the environment in the event of

a. uncontrolled release of materials and/or
b. uncontrolled release of energy

from any particular process.

By hazard we mean the potential to cause harm and this can be measured by the magnitude of the possible consequences.

Consequently, a toxic hazard issuing from a chemical plant will be higher:

- the larger the quantity of toxic material,
- the lower the toxically critical concentration and
- the higher the volatility of the substance.

A hazard caused by an energy release will be higher:

- the larger the amount of energy involved and
- the faster the energy is released.

A major hazard inherent in any particular process can only be regarded as representing a tolerable risk if the probability of it actually happening is very low.

In the three preceding presentations of this workshop on fact finding and basic data it was explained how the hazardous properties of materials are determined experimentally and how hazards can be produced by interaction of hazardous materials with others or with the equipment in a given process.

2. IDENTIFYING THE BASIC HAZARDS

If we want to know the hazards which may be inherent in a chemical process, a process defined by the materials handled and produced, by their quantity and by their intended interaction as given in a process flow

sheet and also by the temperatures, pressures, concentrations and residence times, we must at first check the hazardous properties of the materials handled and establish them by experimental research if they are not
already known.

Eventually, we might come up with a number of different basic hazards,
even for a simple process as shown in FIG. 1.

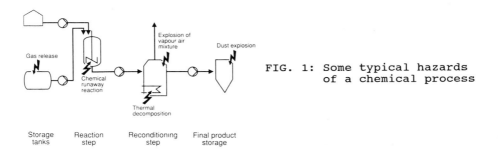

FIG. 1: Some typical hazards
of a chemical process

The first step in identifying these hazards is to compare the hazardous
properties and states of our materials as assessed in numerous experiments with the planned process parameters (in particular: pressure, temperature and concentration). By choosing the appropriate process parameters, we must make sure that none of the basic hazards identified can become real during normal operation.

This goal can be reached by designing a tight plant which prevents volatile material from escaping and by choosing process parameters, which ensure that flammable material cannot be ignited and that possible energy
release from any chemical reaction takes place at such a low rate that it
will always be controlled, e. g. by normal cooling power. This means that
in our process we ought to limit, for instance, temperatures and concentrations to certain fixed maximum values in order to keep the rate of an
undesired reaction near zero.

3. DETAILED IDENTIFICATION OF HAZARDS: FINDING ALL POSSIBLE WAYS THROUGH WHICH A HAZARD COULD BE TRIGGERED IN A REAL PROCESS

3.1   General approach
Any chemical process in a technical plant is subject to different stages
for example

- start up,
- normal operation,
- shut down and
- maintenance of equipment.

A great deal of human intervention is required (possibility of human
error) and the process must also rely on the proper functioning of the
technical equipment which can also fail, as we all know.

The main task in identifying all possible hazards in a chemical process
is therefore to run through all the above-mentioned stages and to systematically search for all the possible ways in which hazards could be
triggered. While doing this, we must anticipate human error and its consequences and we must check foreseeable equipment failures for their consequences. This, of course, is a tremendous task which can only be achieved if a rigourous and very systematic procedure is applied.

It is essential to start hazard analysis in the very early stages of process development because then major hazards are often easily overcome by
a simple change in the process.

Hazard analysis must accompany the whole development and planning procedure for a new process in a new chemical plant and must go into more and
more detail as the process design becomes more detailed. FIG. 2 shows, as
an example, the procedure adopted in my company, Bayer AG of Leverkusen,
Germany.

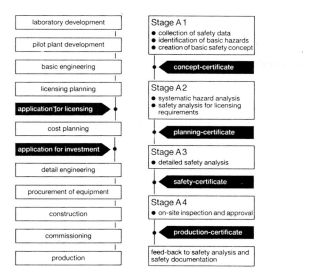

FIG. 2: Systematic analysis procedure

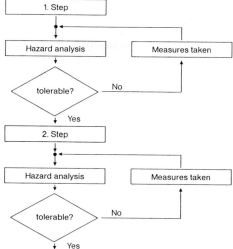

FIG. 3: Stepwise, iterative hazard analysis procedure

As can be seen from the illustration, we have divided safety activities (on the right-hand side of the figure) into 4 stages in increasing detail. The corresponding planning or development stages are shown on the left-hand side of the diagram. At the end of each safety development stage (right-hand side), we require a final check which - upon positive conclusion - results in a certificate. Only with such a certificate at hand may the procedure proceed!

Hazard analysis must be integrated in a stepwise and iterative procedure, alternating between analysis (that is: identification and evaluation of hazard) and synthesis (that is: designing a suitable measure to protect against the hazard).

Since any measure induced against one hazard might create a new hazard - sometimes even somewhere else in the process - our analysis must be an iterative one with backloops before we can proceed to the next step. This is illustrated in FIG. 3.

3.2 Tools and methods available for hazard analysis

The necessary systematic approach in hazard analysis can be reached first by rigourously following the flow sheet of our process from one unit operation to the next or by following chronologically a given procedure. Secondly, it can be backed up by applying certain tools which give assistance.

Table 1 gives an overview over the different methods available for the identification and evaluation of hazards.

TABLE 1.   Systematics of hazard analysis methods

Task	Aim	Working principle	Method
1) Identification of hazards	1) To complete the safety concept	1) Memory-jogging	1) Use of check lists
			2) Relationship charts
		2) Use of search aids and tables	3) Failure modes and effects analysis
			4) Action error analysis
			5) Hazard and operability studies
2) Evaluation of hazards according to probability of occurrence	2) To optimize safety systems with regard to reliability and availability	3) Representation of inter-connections of failures in graph form and evaluation of probability	6) Incident sequence analysis (inductive)
			7) Fault tree analysis (deductive)
3) Evaluation of hazards according to their consequences	3) To minimize the hazard potential and devise optimum protective measures	4) Mathematical analysis of physico-chemical processes	8) Hazard consequence analysis

The table lists five different methods which can be used to identify hazards in chemical processes and plant. These basically rely on two different working principles. It also gives two methods of evaluating hazards in terms of probability of occurrence and one for the evaluation of hazards in terms of possible consequences.

What can the identification methods do for us?

Many types of hazards and their sources are quite well known from experience. Such experience can be written down systematically and then used as a **checklist** for hazard identification. Consider the stirring vessel in FIG. 4 which could be used for carrying out an exothermic chemical reaction.

In a checklist for the prevention of a runaway reaction we would certainly be asked to consider possible hazards resulting from:

- loss of power supply, (stirrer stopps)
- loss of cooling water,
- excess of reaction initiator,

and in our analysis we would then have to investigate the question of whether or not this might cause a hazard. Such questionning might quite normally initiate additional experimental investigation.

In order to ensure that the right questions are always asked in such case, without forgetting a single one, at Bayer we have, drawn up extensive checklists for what we call "hazard areas". One such checklist covers one type of hazard, e. g. a runaway reaction, and asks for information on all problems that may trigger the hazard.

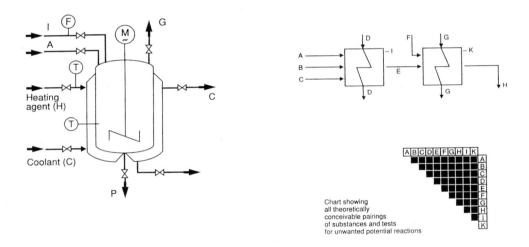

Chart showing
all theoretically
conceivable pairings
of substances and tests
for unwanted potential reactions

FIG. 4: Reaction vessel with stirrer   FIG. 5: Use of relationship chart

A checklist can be considered to be a memory-jogger. The same is true for another technique used to identify e. g. reaction hazards, the use of **relationship charts** or **matrices**. If we want for instance (see FIG. 5) to find out about all hazardous reactions between the materials in a two-step chemical reaction process, we can immediately point out all theoretically conceivable binary interactions in a two-dimensional matrix of all substances present in our process.

Some of the chemical interactions shown in the matrix of FIG. 5 have no physical relevance, some do not pose a threat and some might need further investigation. Here again, the formal approach will help to avoid overlooking a problem.

Keeping in mind that in a well designed process and plant a hazard can only arise as a result of deviation from normal conditions, that is by an error, a fault or a failure, systematic investigation of all imaginable deviations from

- correct functioning of equipment,
- required human action or
- relevant process parameters and functions

and their consequences is a very helpful procedure.

Such a procedure is applied in three rather similar analysis methods using tables and guide words to help identify deviations, i.e. in

- failure mode and effects analysis (FMEA)
- human action error analysis
- hazard and operability studies (HAZOP).

In FMEA all hazards originating from the malfunction of equipment are systematically identified by assuming that the equipment under investigation does <u>not</u> function as it should.

In human action error analysis guide words for human errors like: "too late", "wrong" (direction, object) are applied to the action required in order to identify and analyze hazards originating from such errors.

Finally, in HAZOP, guide words like "less of ..." or "more of ...", "reverse" or "none" are applied to process parameters like temperature, pressure, residence time or flow and the consequences are analyzed in order to detect hazards. Table 2 gives an example of how the methods can be applied to our process in the reaction vessel as shown in FIG. 4. The procedure is self-explanatory and systematically follows the flow-diagram of the process.

TABLE 2.   Example of hazard analysis

Method	Deviation	Cause	Consequences	Safety measure
Failure modes and effects analysis (failure of equipment)	Stirrer: does not turn, turns wrong way:	No electricity, motor defective, installed wrongly	Heat transfer impeded, solids precipitate	Revolution counter, feed shut-off, control unit
	Steam valve: does not close	Jammed	Temperature increase in reactor	Feed shut-off
Action error analysis (human error)	"Too late": steam off, cooling water in	Operator not paying attention, overworked	Temperature increase, danger of uncontrolled reaction	Alarm, feed shut-off, emergency cooling
	"Not": steam off	Operator not paying attention		Automatic shut-off
	"Incorrect": product	Mix-up	Unknown reaction	Labelling, analysis
Hazard and operability studies (deviations in functions and process variables)	"More": steam, heat	Pressure reduction fails, no cooling	Temperature increase, danger of uncontrolled reaction	Shut-off of steam by hand, feed shut-off, emergency cooling
	"None": cooling water, stirring	Valve does not open, driving mechanism defective	Heat dissipation impeded, temperature increase	Feed shut-off, emergency cooling
	"Reverse": direction of stirrer	Installed wrongly	Solids precipitate	Direction indicator, control unit

Everyone attending this IUPAC-Workshop will certainly use one of the methods described during the practical exercise on day two. It will then become clear that the procedures can be very tedious and that, even so, they cannot guarantee that every hazard will be detected. Rational reasoning, good knowledge of the process and common sense are therefore a "must".

As far as methods for the evaluation of the probability and consequences of identified hazards are concerned, reference is made to paper mentioned in the appendix. The relevant methods are specialist methods and proper application poses quite a few problems.

At Bayer hazard identification relies entirely on experimental investigation of the materials involved in our process for basic factfinding, and on checklists and a modified HAZOP procedure for the analysis of process hazards.

4. THE RANKING OF HAZARD IDENTIFICATION AND ANALYSIS IN THE INTEGRAL SAFETY TASK

The principal goal of hazard analysis is - of course - to provide adequate safety measures for a given process and plant.

During a real planning procedure the tasks of analysis and constructive measures are closely linked as shown earlier; but for didactic purposes, this conference has spread out the different tasks involved in "identification of hazards" on one hand and "measures for risk reduction" via lowering the probability of hazardous incidents and mitigation of consequences on the other hand.

In order to show the coordination and inter-relationship as well as the individual importance of the different safety tasks, I should like to enumerate once again what duties have to be performed to ensure a safe chemical process in an industrial plant.

We have to:

1. minimize and inactivate the hazard potential of our process and

2. avoid activation of the residual hazard potential in case of errors, faults and failures

This can be achieved by:

1 a. identifying all possible basic hazards via experimental testing of our materials

1 b. evaluating these hazards by means of mass and energy balances

1 c. minimizing the hazards by substituting other materials or minimizing the quantity used in our process

1 d. attenuating or inactivating the residual hazard potential by setting safe process and plant parameters

and

2 a. identifying all possible ways of activating a potential hazard via systematic hazard analysis

2 b. evaluating the possibilities of activation in terms of their probability

2 c. minimizing possibilities of faults and failures in our plant by

- suitable design and layout
- quality assurance for equipment
- providing multiple barriers

2 d. designing and building the plant so as to tolerate failures by

- resilient design
- application of the fail-safe principle
- provision of multiple barriers

Four of these eight tasks are analytical (namely 1 a, 1 b, 2 a and 2 b) and four are matters for creative engineering (1 c, 1 d, 2 c, 2 d).

Hazard identification comprises two of these eight tasks, but it is very fundamental.

Only if hazard identification is applied at all stages in the life of processes and plants and to every necessary operation, can a safe process and plant be ensured.

LITERATURE

/1/   "Safety analyses for the systematic checking of chemical plant and processes - Methods, benefits and limitations" by V. Pilz, German Chem. Engng. 9 (1986) 74/83 (in English).

Analysis and assessment of hazards and risks

J L Hawksley

ICI Group Safety and Environment Department

Abstract – Safety in chemicals production requires, firstly, an understanding of the "hazards". It is then the responsibility of those managing the Business to ensure that appropriate measures are taken to make the likelihood of harm actually occurring, that is the "risk", as low as is acceptable or tolerable. The understanding of the hazards and the likelihood of harm arising requires technical "analysis" of the possible harmful consequences and the likelihood of events that could give rise to those consequences. The "assessment" of the acceptability or tolerability of the resulting risks requires value judgements to be made. Hence, there is a progression from hazard analysis through risk assessment. This paper outlines and discusses some of the various techniques that can help with the analysis and assessment; these range from the mainly qualitative to those with varying degrees of quantification.

INTRODUCTION

Assessing risk is an essential part of managing the production of chemicals to ensure safety by proper planning, good design, sensible location and careful operation of process installations. There are many approaches varying from the essentially qualitative to those that are more quantitative. In all cases, informed judgement based on experience is an essential ingredient. This paper outlines some approaches relevant to considering potential risks from chemical process accidents with the potential to cause acute harm. But it is not a review of all. Neither does it consider other areas of risk assessment that are nevertheless important, for instance, consideration of the chronic risks from low level, long term exposure to chemicals.

First it is best to be clear about what is meant by some of the terms. "Hazard" and "risk" are often used indiscriminately, but in the context of some of the dangers that might arise from chemical processes are taken as:

Hazard – a situation with the potential to cause harm.
Risk – the probability of harm arising.

"Analysis" and "assessment" are also often interchanged but, in this discussion, "analysis" refers to the technical process of identifying, understanding and evaluating hazards and risks. "Assessment" refers to the value judgements that have to be made when considering the significance of a situation, eg the acceptability of a risk. Before an "assessment" can be made some "analysis" is required; hence "hazard analysis" is a precursor to "risk assessment".

NB – the term risk assessment is also used, and might be defined differently, in other contexts.

GENERAL PROCEDURE AND PRACTICE

The general risk assessment procedure when considering potential accidents that might arise during chemicals production has a number of elements:

1. Identification of the sources of hazard and the potential hazardous events (accidents).

2. Analysis of the mechanisms by which identified hazardous events could occur.

3. Evaluation of the consequences of the identified hazardous events (consequence here refers to a potentially harmful condition measured in terms of, for example, toxic gas concentration, thermal radiation, explosion overpressure).

4. Evaluation of the likelihood of occurrence of the identified hazardous events.

5. Evaluation of the potential risk from the hazardous events (risk here refers to the likelihood of a stated level of harm or damage arising and is a combination of the frequency and consequence of the hazardous event(s) taking account of the probability of exposure (say of a person) to the hazard).

6. Consideration of the degree of risk.

Elements 1 to 3 constitute "hazard analysis" which becomes "risk anlaysis" when the frequency and probability considerations in elements 4 and 5 are added. The risk analysis then becomes "risk assessment" when the value judgements in element 6 are embraced.

In practise the various elements may be mainly qualitative or may involve varying degrees of quantification. The term "quantified risk assessment" (QRA) is used loosely to describe the procedure when numerical estimates or calculated values are used in elements 3, 4 and 5 and judgements in element 6 made with reference to numerical guidelines or criteria.

There is a considerable number of techniques and tools to use at various stages of the process. These range from experimental techniques, with which to identify hazards on an empirical basis and generate data for subsequent analysis, to analytical techniques (HAZOP, fault tree and event tree analysis, etc) to aid the thought processes. Some of the techniques are outlined in other Workshop papers (see also ref. 1 and 2).

The elements of the risk assessment procedure are used to generate information to aid decision making. For these purposes the guiding principle is that the procedure needs only to be taken as far as is necessary to be helpful in making the decision. In general it can be said that element 1 is the most important; to achieve safety it is vital to identify hazards and potential accidents thoroughly. Then the need (and ability) to address the other elements varies from one situation to another and this determines how the procedure is applied. For instance, in many cases having identified a hazard in element 1 a decision based on experienced judgement may well be that the risk from the hazard will be made sufficiently low by applying an accepted code or standard, ie elements 2 to 5 are by-passed. In other cases consideration of some or all of those elements will be necessary. Just as the extent of application of the various elements varies, so does the extent of quantification because the confidence with which calculations of consequence, frequency and, particularly, risk can be carried out varies widely for different situations and the results may be of limited accuracy.

The differing extents of application of the elements of risk assessment are illustrated by some areas of decision making for which quantification of more or less of the elements can be helpful, for example:

a) Ranking hazards to determine priorities.
b) Choosing between alternatives (eg process routes, safety systems).
c) Selecting certain design parameters (eg. for reliability of shutdown or protective systems).
d) Deciding acceptability (eg. how safe is safe?).

Significant points are that a) and b) are essentially comparative exercises and as such do not require all the elements to be addressed. Also for comparative assessments, any uncertainties in the quantifications are less of a drawback. For c) and d) however, more absolute judgements have to be made and so some quantitative guidelines are required. For c) these may be defined in terms of event frequencies or consequence, but for d) guidelines of acceptable levels of risk are necessary if the procedure is to be of real help. For assessments that require absolute judgements, uncertainties in the quantifications limit the extent to which the procedure can realistically and usefully be applied.

HAZARD ANALYSIS

Basic data and indicative criteria

Having identified a hazard the basis for a measure of hazard is a consideration of the appropriate "intrinsic" properties of the substances themselves and "extrinsic" factors with respect to the containment of the substance and its location, both of which may have a bearing on the degree of hazard. For instance:

1. "Intrinsic" properties of substances
    - toxicity (OEL, control limits, LC_{50}, etc)
    - flammability (flashpoint, flammability limits, heat of combustion, etc)
    - thermal stability
    - reactivity
    - explosibility

- volatility (boiling point, vapour pressure)
- dispersibility (mol wt, vapour density)

2. "Extrinsic" factors of containment/location
   - temperature (relative to ambient boiling point)
   - pressure
   - environment (with respect to its effect on integrity of containment and consequences of loss of containment)
   - inventory

A judgement of the degree of hazard can be made based on an awareness of the extent to which hazard increases or decreases with such factors, eg vapours heavier than air will disperse less readily than those lighter than air, flammable liquids processed at pressure and above their ambient boiling point will "flash" if loss of containment occurs with more chance of forming a potentially explosive vapour cloud.

Such data can be used together with "indicative" criteria which define substances as "very toxic", "toxic", "highly flammable", "flammable", etc. In some cases the indicative criteria may be statutory, see for example ref. 3 which also defines inventories for specified dangerous substances which, when exceeded, constitute so-called "Major Hazards".

Table 1 illustrates some of the relevant data for several toxic gases.

TABLE 1: Toxic gases – examples of some of the data needed for consideration of relative hazards

Substance	OEL (TLV) ppm	Toxic concentrations/ppm[1]		Boiling point $^{\circ}$C	Vapour density relative to air	"Major Hazard" inventory (Ref 3)	Substance Hazard index (SHI)(Ref 6)
		Causes severe irritation in humans (eyes/nose/throat)	Lowest LC_{50} (animal tests)				
Ammonia	25	400	1066	-33	0.59	500 te	7068
Chlorine	1	15	293	-34.5	2.49	25 te	335526
Phosgene	0.1	3	110	8.3	3.41	750 kg	789474
MIC	0.01	2	5	39	1.97	150 kg	22668

1. A selection, only, of data taken for illustration from convenient sources, eg Sax "Dangerous Properties of Industrial Materials" and CIA Codes of Practice for ammonia and chlorine.

HAZARD INDICES

For flammable substances an objective measure of hazard can be made using the Dow Fire and Explosion Index (F&EI) (ref. 4). This can be applied to sections of a process (eg reaction, compression, etc) to give a ranking based on the properties of the materials present, the quantity, operating conditions and type of process. The ranking has a quantitative basis and is built up from numbers generated by applying conversion factors to the various basic data. The numbers in themselves have no physical meaning, but the higher the number the higher the hazard. The Mond Fire and Explosion Index (ref. 5) is a development of the Dow Index with the calculated index converted into qualitative descriptions on an eight point scale ranging from MILD through HIGH to VERY EXTREME. The conversion factors have been chosen by considering many units of recognised hazard.

With the Mond Index, as well as generating an overall hazard index, separate indices can be calculated to identify which particular hazard or hazards (eg fire, internal explosion, aerial cloud explosion) contribute most significantly to the overall hazard.

Table 2 illustrates a comparison of different process routes to the same product showing a possible new route (No 2) as potentially more hazardous than that in current use (No 1).

Table

TABLE 2: Example of Mond Fire and Explosion Index assessment of the hazards of the main units of alternative routes to the same product

	Unit	Fire	Internal Explosion	Aerial Explosion	Overall Risk
Route 1	1	High	Moderate	Light	Moderate
	2	Moderate	Low	Light	Low
	3	Moderate	Low	Light	Low
	4	High	Moderate	Light	Moderate
	5	High	Moderate	Light	Low
	6	High	Moderate	Light	Moderate
	7	Moderate	Low	Light	Moderate
	8	Moderate	Low	Light	Low
	9	Moderate	Low	Light	Moderate
Route 2	1	Intensive	Moderate	Very High	Extreme
	2	Moderate	Moderate	Very High	High
	3	Light	Low	High	Moderate
	4	Light	High	Low	Moderate
	5	Light	Low	Low	Moderate
	6	Moderate	Low	Low	Moderate
	7	High	Low	Light	Moderate

Both indices can be extended to give some measure of risk. With the Dow Index a procedure is available to estimate maximum probable property damage ranging from a "base" value (worst case) to an "actual" value which assumes reasonable, but not perfect functioning of protective measures. With the Mond Index account can be taken of the various extrinsic safety measures that are or would be used to ensure safe operation. These "off setting" factors are usually found to reduce the ranking by one or two levels. Experience indicates that at a hazard ranking up to HIGH safe operation is generally possible without particularly specialised extrinsic protection. Above HIGH increasingly complex extrinsic protection is required.

For toxic substances there are not, at present, such well developed indices; perhaps not surprisingly, given the wide range of toxic substances and effects. In the US a Substance Hazard Index (SHI) has been proposed as a basis for identifying substances requiring more stringent statutory control (ref. 6).

SHI = EVC / ATC

Where EVC is the equilibrium vapour pressure at 20°C (atm.10^6) and ATC is the acute toxic concentration (ppm) defined as the lowest reported concentration judged likely to cause death or permanent injury to humans after a single exposure of one hour or less. Above a value of 5000 substances would require special attention; example values for some toxic gases are shown in Table 1.

Dow have developed a Chemical Exposure Index (CEI) to provide a method of rating the relative potential of acute health hazard to people from possible chemical release incidents (ref. 7). The CEI is built up from five factors which influence the magnitude of exposure from any potential release, ie the acute health hazard factor (similar to substance hazard index), the quantity volatilised on release, the distance to locations where people could be exposed, the molecular weight (which relates to vapour density and dispersion) and processvariables (temperature, pressure, reactivity). Each factor is assigned a value 0 to 4 (0 to 5 for acute health hazard) using guidelines. The CEI is thus a number on the scale 0 to 1280. A typical value is 500 for a large (≈2000 tes) chlorine storage and processing installation. The index is not intended to define situations as acceptable or otherwise, but to focus attention on the level of attention, eg the frequency of process reviews and audits.

Thomas (ref. 8) investigated the possibility of a toxic hazard index similar to the Dow and Mond Fire and Explosion Indices. The results showed some promise, but the procedure was involved and has not found favour. Its basis was the relative area that could be affected by the release of toxic material (see next section).

Consequence Analysis

A large amount of information exists to allow estimates of the consequences of hazardous events involving the release of toxic or flammable substances or the sudden release of energy. Estimates can be made either by comparison with actual incidents or tests or by calculation (see for example refs. 9, 10 and 11). Although calculation methods are almost continuously being refined many assumptions and simplifications are often required, so it must be appreciated that the accuracy of the results may be limited.

The general procedure is first to define the "source term", for example, the quantity or rate of material released. Then calculations of the hazardous affect are carried out using appropriate methods and criteria so that, for example, the "hazard range" to a defined level of harm can be estimated (see Figure 1). Table 3 gives some typical criteria (see also ref. 12).

TABLE 3: Typical criteria used to determine the extent of harm arising from hazardous events

	1. High probability of fatality	2. Threshold of fatality	3. Threshold of injury
Explosion (overpressure)	0.3 - 0.5 bar	0.1 - 0.2 bar	0.03 - 0.06 bar
BLEVE/ fireball	450 - 600 kJ/m^2	200 - 300 kJ/m^2	50 - 100 kJ/m^2
Toxic vapour cloud	LC_{50}*	LC_{lo}*	IDLH*

* LC_{50} – concentration likely to cause death of 50% of those exposed
LC_{lo} – concentration likely to cause death of a small proportion of those exposed
IDLH – concentration "immediately dangerous to life and health"

Determining the source term involves defining the failure that occurs (say, the fracture of a branch pipe or a vessel) and then carrying out the appropriate fluid flow calculation for single or two phase flow as the case may be. The flow regime has an important bearing on the hazard; for a given pressure and hole size the mass rate of release will be approximately four times less if two phase rather than liquid only flow occurs and approximately a further four times less if the flow were vapour only.

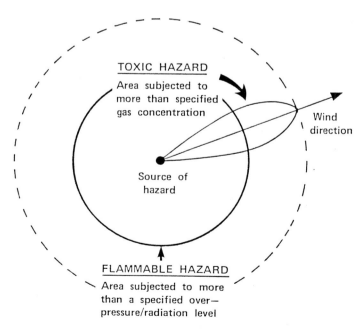

Fig. 1 Consequence analysis - hazard range and area affected

The source term calculations may also have to allow for vaporisation of discharged liquid which might be immediate "flash evaporation" or subsequent evaporation from an accumulated pool. The pattern may change with time and include assumptions with respect to action being taken to stop the flow. Figure 2 shows an example.

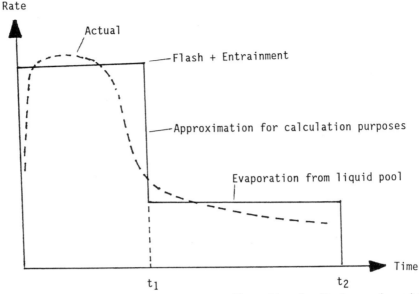

t_1 = time to stop spill or time to discharge inventory
t_2 = time to evaporate all of pool

Fig. 2   Source term - pattern of vapour evolution of a spill of
highly volatile liquid (NB not to scale)

Having defined the source term the subsequent analysis depends principally on whether the hazard is flammable or toxic.

Take the release of a flammable vapour. Several outcomes may be possible; for example, if ignition was rapid a fire might be the result, if ignition was delayed an explosion could be more likely. Event tree analysis may help decide which outcome is the more likely (see Figure 3).

Where fire is the hazardous event, whether jet fire, pool fire or fireball, methods are available to predict the thermal radiation flux or dose that could occur at various distances from the source of hazard. Where explosion is the hazardous event, methods are available to predict, from the energy likely to be released, the overpressures that could be generated by the shock wave. In either case "hazard ranges" can be estimated.

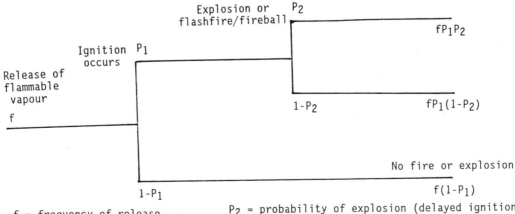

f = frequency of release P_2 = probability of explosion (delayed ignition)
P_1 = probability of ignition

Fig. 3   Event tree showing different outcomes from release of
flammable vapour

For the release of a toxic vapour, the analysis involves gas dispersion calculations to predict the concentrations of gas that could occur at various distances from the source of hazard. Much work has been done in recent years to refine and validate gas dispersion models particularly so-called "dense gas" models which must be used when the vapour is heavier than air. However, many assumptions have to be made and the models tend to be valid only for "ideal", eg open, terrain rather than "real" situations where, for instance, obstructions in the form of building and trees will affect the dispersion that actually occurs. Gas dispersion is very dependent on both wind speed and weather conditions (atmospheric stability) which can make it difficult to specify maximum hazard ranges (see Figure 4).

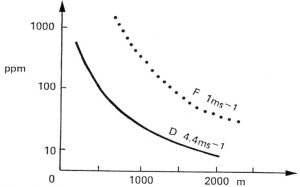

Fig. 4 Estimates of the variation of toxic gas concentration with distance from point of release for two sets of atmospheric stability and wind speed (F inversion, D neutral)

Gas dispersion calculations provide an approach to the objective ranking of the relative hazard of different toxic gases by calculating, for given sets of circumstances, the areas that might be affected by gas concentrations capable of similar hazardous effects. Table 4 shows the results of such calculations for the release of 1kg/s of different toxic gases. Similarly, by varying the circumstances (eg. the quantity of gas) situations notionally having the same degree of hazard (similar area affected) can be defined (see example for phosgene/chlorine in Table 4).

For the purposes of this illustration the calculations were done, for simplicity, using readily available toxicity data. These are not necessarily strictly comparable, but do illustrate a major problem, when considering toxic gas hazards, which is in specifying what concentration (or dose) will have what effect. Most of the data on lethal doses derive from animal tests which may have doubtful relevance to the effects on humans. For instance, the animal data might suggest that MIC is very significantly more hazardous than the other gases which would seem not to differ greatly. The data from human exposure gives a different picture with more significant differences between ammonia, chlorine and phosgene. MIC still shows as the most hazardous, but less markedly so. toxicity data has to be used with caution and with expert advice. Reference 13 is an example of an attempt to make best estimates of the toxic effects of chlorine from the data available.

TABLE 4: Estimates of areas within which gas concentration exceeds a given value - a possible basis for comparison of toxic gas hazards

Basis — 1kg/s emission at 20°C
  — wind 4.4m/s
  — weather category D

Toxic gas	Area of isopleth/hectares ($10^4 m^2$)	
	Severe irritation (humans)	Lowest LC_{50} (animals)
Ammonia	0.44	0.14
Chlorine	5.1	0.26
Phosgene	18.3 (5.1*)	0.39
MIC	47.6	18.6

NB: These illustrative calculations were performed for the toxicity values given in Table 1. They do not derive from a critical review of all available data, hence the figures cannot be taken as a definitive indication of relative hazards.

* Area for 0.3kg/s emission rate

RISK ASSESSMENT

Without quantification of probability

In practise most assessments are carried through without quantification of probability, relying on informed and experienced judgement to decide what is likely or unlikely depending on circumstances. This approach is most robust when the worst events that need to be taken into account can reasonably be specified and the relevant consequences and "hazard ranges" determined. It can then be said with some confidence that the risk (of whatever level of harm the hazard range is determined for) will be very low at locations beyond the hazard range. A generalisation is that this approach is easier to apply the less severe are the potential consequences. Also, it is easier the less the variation in possible consequences for a given hazardous event. In this respect it can be applied to flammable hazards perhaps more readily than toxic gas hazards. For emissions of toxic gas the large variation in consequence with different weather conditions (see Figure 4) can make it difficult to specify realistic maximum hazard ranges without specific consideration of the probability of the more severe cases.

With quantification of probability

In some cases simply quantifying the likelihood of the identified hazardous events occurring and/or the probability of a particular consequence can be helpful. But, in the limit, a quantified risk assessment involves quantification of a specified risk of harm occurring. Commonly either or both of two expressions of risks are used:

Individual risk – the chance of a person at a given location being killed, expressed typically as chances per year.

Societal risk – the likelihood of accidents causing multiple fatalities expressed as the frequencies of occurrence of, say, 10, 100 ... deaths.

The basic equation for computing individual risk, r, is straightforward:

$$r = HP_1 \cdot P_2 \cdot P_3 \tag{1}$$

Where H is the frequency of occurrence of the hazardous event (the "Hazard Rate") (y^{-1}), P_1 is the probability that the location in question would be affected by the hazardous event, P_2 is the probability that a person would be present in the area affected and P_3 the probability that fatal injury would occur as a result of the harmful condition. In some cases the probabilities can be evaluated separately. In others the probabilities may not necessarily be independent, eg both P_2 and P_3 may depend on P_1 and it may be easier to evaluate combinations. There may be other probabilities within the basic ones, eg in some cases P_2 may include an allowance for the probability of escape from the affected area.

Example – A reactor within an enclosed building was required for a new process for which the existing vent would be inadequate. Additional procedural and instrumented protective systems were proposed to prevent overpressure. The systems were analysed and fault trees derived from which the likelihood of failure of the reactor due to overpressure was estimated to be 2×10^{-4}/y. Failure of the reactor was judged likely to affect the whole of the enclosure ($P_1 = 1.0$) with a high probability of fatal injury to anyone present ($P_3 = 1.0$). An operator would be present in the enclosure for 30 minutes each day ($P_2 = 30/24 \times 60 = 0.02$). The risk to the operator was estimated to be $2 \times 10^{-4} \times 1.0 \times 0.02 \times 1.0 = \underline{4 \times 10^{-6}}$/y.

In any risk analysis it is important to look out for so-called "common mode" effects which have the effect of making a combination of events perhaps more likely than expected. For instance, in the above example, if the presence of the operator would be instrumental in causing the overpressure to occur then P_2 would be 1.0 not 0.02.

In other circumstances, eg the consideration of the risk at off-site locations from possible releases of toxic gas, determining the probability of locations being affected can be more complex. Also it may require the contributions from a number of failure cases to be considered and summed. For each failure case and various combinations of weather conditions (ie wind speed and atmospheric stability) the size of plume within which gas concentrations would be sufficient to cause the specified harmful effect is estimated.

Figure 5 illustrates one relatively simple approach in which the probability of the wind being in any direction is assumed uniform around 360°. The probability of the location X at a distance L from the source of hazard being affected is then proportional to the plume width W, at the location and the probability, Pw, of the weather condition giving that particular plume size.

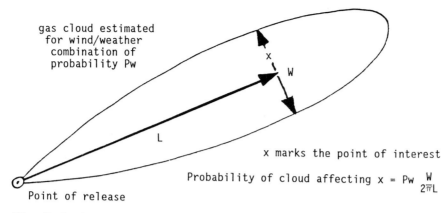

Fig. 5 Risk analysis - estimating the probability of a gas cloud
affecting a given location

This procedure is repeated for a sufficient number (m) of combinations of weather conditions
to reflect the range that, from a knowledge of meteorological data for the locality in
question, could be foreseen. Hence the total probability of location X being affected for
one given failure case is:

$$P_2 = \sum_{i=1}^{m} (P_w^{(i)} . W_i / 2\pi L) \tag{2}$$

Typically m ranges from 2 to 25 depending on the data available and the precision being
sought. A first analysis may assume that a person may always be present at the location
($P_1 = 1.0$) and P_3 will depend on the gas concentration for which the plume was calculated.

The procedure is repeated for the number of failure cases (n) that could affect the
location and total risk is:

$$r = \sum_{j=i}^{n} (H_j . P_1 . P_2^{j} . P_3) \tag{3}$$

By repeating this procedure for locations at different distances from a source of hazard a
"risk profile" (see Figure 6) can be determined or "risk contours" drawn around an
installation linking points of equal risk (see for example refs. 14 and 15).

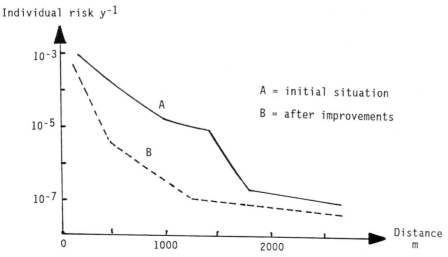

Fig. 6 Examples of risk profiles showing variation of risk with
distance from source of hazard

Various refinements can be made. For example, to allow for the variability of wind direction is relatively simple. More difficult is to allow for the infiltration of gas into buildings which results in "indoor risk" being different (mostly lower) than "outdoor risk". Understanding of this aspect is important for developing the most effective emergency action which is, in many cases, to seek shelter (evacuation may not be necessary, or desirable, unless the emission is prolonged). (See ref. 16).

From this brief outline it can be appreciated that some risk analysis can be a complex procedure, in many cases requiring computer methods to handle the many combinations of events and circumstances that may have to be considered. The procedure is subject to many uncertainties which limit the accuracy of the results to an uncertainty of <u>at best</u> half to one order of magnitude.

These limitations have been discussed elsewhere (see refs. 17, 18 and 19). Uncertainties in predictions of consequences of accidental events have been already mentioned. Predicting the size and frequency of accidental releases is another area of significant uncertainty. Often analyses centre around low probability events of high potential consequence and, in the limit, severe failures of pipes and vessels are dominant, but the likely size of hole and its frequency of occurrence can not be predicted with any precision.

The analysis results in a numerical prediction of risk and so some numerical criteria or guidelines are required to guide the assessment. There are no universally accepted criteria to determine what risks are acceptable or tolerable. Rather there is a band within which judgements have to be made taking account of many other factors, for instance, the benefits that might offset the risk. The Royal Society Study Group on Risk Assessment (ref. 20) is often quoted as an authoritative view.

They defined a range of indicators. At one extreme they noted that the imposition of an annual risk of death to an individual of 10^{-3} might not be "totally unacceptable if the individual knows of the situation and enjoys some commensurate benefit and everything reasonable has been done to reduce the risk". At the other extreme, they judged that an additional annual risk of death to an individual from some activity of 10^{-6} (ie a one in a million chance per year) would commonly be regarded as trivial, adding that it might, in some circumstances be ten times less (10^{-7}) and in others ten times as great (10^{-5}).

A number of specific average risks (from UK data) within that range are (refs. 21 and 22):

	Risk of death per person per year
Deep sea fishing	880×10^{-6}
Quarrying	390×10^{-6}
Coal extraction	106×10^{-6}
Road accidents	100×10^{-6}
Accidents to adults in the home	50×10^{-6}
Accidents at work (average)	30×10^{-6}
Chemical industry	27×10^{-6}
Work in offices/shops	5×10^{-6}
Lightning strike	0.1×10^{-6}

(The risks for work activities are averages for the population "at work", others are averages for total population).

In assessing workplace risks from specific hazards it is relevant to note the average achieved and to aim lower. On that basis the risk from the reactor in the example above might be regarded as tolerable. Given a "target" risk that should not be exceeded the risk equation (1) can be rearranged to derive a "target" hazard rate from which the necessary reliability of protective systems can be determined.

It is generally felt that risks to the general public from a hazardous activity should be less than the risk that might be tolerated by those engaged in the activity. The "one in a million" (1×10^{-6}/y) risk is often taken to be the general order of tolerability. Some authorities have proposed criteria that risks to the public should, in particular circumstances, not exceed 10^{-5} (UK), 10^{-6} (Netherlands) and levels below which risks can be regarded as insignificant, 10^{-6} to 3×10^{-7} (UK), 10^{-8} (Netherlands) (see refs. 23 and 24). The differences are, in part, due to different ways of predicting the risks to compare against the criteria. When using any criteria in an assessment it is important that the underlying assumptions are understood.

It is also important not to lose sight of the inherent uncertainties in quantifications of risk and make judgements on precise comparisons of predictions with criteria. Often the technique is best used in a comparative rather than absolute mode. In Figure 6 for example the change that alters the risk from A to B could be regarded as a significant worthwhile improvement without regard to the actual value of the risks predicted.

Assessments of "societal risk" are more problematical. The analysis requires the
estimated area affected by a given hazardous event to be combined with estimated numbers of
people within that area to give an estimate of total casualties. By considering the range
of events possible and their possible outcomes so-called FN curves can be generated (see
Figure 7).

This approach is, in principle, particularly relevant where individual risk is shown to be
low, but the possibility of events with very severe consequences has to be considered; an
example is some transport hazards. However, predictions of societal risks are subject, in
general, to even more uncertainties than individual risk, particularly when it comes to the
extremes of consequence (high N) at very low frequency (low F). Also, there is even more
argument and less agreement over criteria against which to assess predictions.

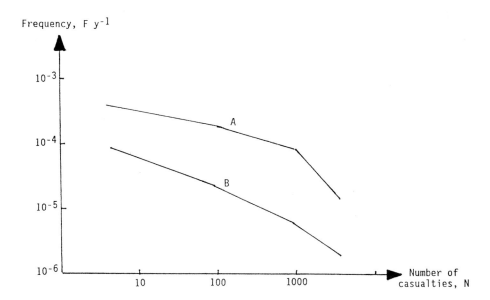

Fig. 7   Examples of FN curves to indicate the frequency, F, of
accidents causing N or more casualties for alternative
options A and B

One study (ref. 25) of a number of examples of the use of risk assessment in decision
making concluded that consistent criteria against which to judge FN predictions could not be
set. The contribution of the predictions to the decision making process mostly came from
the display of the relative risks of different options (eg A and B on Figure 7) rather than
the actual values estimated. However, there will undoubtedly be further development of
societal risk assessment.

CONCLUDING COMMENTS

A number of approaches to hazard and risk analysis and assessment have been outlined. More
detail and discussion of these and other methods will be found in the references quoted.
There is no single best method. Different ones applied to differing extents are appropriate
in different circumstances.

A considered and selective use of the appropriate methods contributes to achieving safety in
chemicals production primarily as a result of the systematic identification and logical
analysis of hazards. The quantification of consequence and/or frequency, and, in the limit,
risk, can in some cases promote a better appreciation of the hazards and help decide the
stringency of the control methods necessary to limit the risk.

A key point is that the methods are all "tools" to aid decision making and the weight that
is given to any quantitative analysis in making an assessment must bear in mind the
uncertainty of the data used and the numbers generated.

A further key point is that there will always be assumptions underlying any assessment.
Most important is the assumption that the processes and plants assessed are or will be
operated in accordance with good management practise - for example, safety critical
equipment will be inspected, tested, maintained, etc, to ensure its continual integrity.
Hence, the various methods of hazard and risk analysis and assessment cannot show that a

process or plant _is_ safe. They can help demonstrate that it _can_ be safe, but safety is ultimately dependent on management control and the systems and procedures set up to implement that control.

REFERENCES

1. "Methodologies for Hazard and Risk Assessment in the Petroleum Refining and Storage Industry", Report 10/82, CONCAWE, den Haag (1982).

2. T A Kletz "HAZOP and HAZAN - Notes on the Identification and Assessment of Hazards", IChemE (1983).

3. European Council directive on the Major Accident Hazards of Certain Industrial Activities, (82/501/EEC) (1982).

4. Chemical Engineering Progress Technical Manual "The Dow Fire and Explosion Index", AIChE, New York.

5. The Mond Index, details available from ICI Chemicals and Polymers Ltd, Explosion Hazards Laboratory, Winnington, Cheshire, England.

6. OSHA Proposed Rules "Process Safety Management of Highly Hazardous Chemicals", Federal Register, 55, 137, 17 July, Washington DC, (1990).

7. E L De Graaf, "Dow Chemical Exposure Index", 6th International Symposium Loss Prevention, Oslo, 19-22 June (1989).

8. A R Thomas, "An Acute Toxicity Index for Industrial Materials, MSc Thesis, UMIST (1988).

9. F P Lees, "Loss Prevention in the Process Industries", Butterworths, London (1980).

10. "Manual of Industrial Hazard Assessment Techniques", Office of Env. and Sci. Affairs, The World Bank (1985).

11. "Guidelines for Chemical Process Quantitative Risk Analysis", Center for Chemical Process Safety, 585pp, New York (1989).

12. J L Hawksley and R Grollier-Baron, "Some Aspects of Risk Management in the Petrochemical and Process Industry", IBC Conference, Safety and Reliability Objectives in High Technology and Industry, Amsterdam 2-3 March (1989).

13. "Chlorine Toxicity Monograph", IChemE (1987).

14. R P Pape and C Nussey, IChemE Symposium Series No 93, p367 (1985).

15. R M Pitbaldo, R J Shaw and G Stevens, "The SAFETI Risk Assessment Package and Case Study Application", International Conference on Safety and Loss Prevention in the Chemical and Oil Processing Industries, Singapore (1989).

16. D J Wilton and B W Zelt, "The Influence of Non-Linear Human Response to Toxic Gases on the Protection Afforded by Sheltering-in-Place", OECD/UNEP Workshop on Emergency Preparedness and Response, Boston, 7-10 May (1990).

17. "CEFIC Views on the Quantitative Assessment of Risks from Installations in the Chemical Industry", European Council of Chemical Manufacturers Federations (1986).

18. "Risk Analysis in the Process Industries, Report of the International Study Group on Risk Analysis", European Federation of Chemical Engineering Publication Series No 45 (1985).

19. J L Hawksley, "Risk Assessment and Project Development", The Safety Practitioner, 5, (10), 10-16 (1987).

20. "Risk Assessment - A Study Group Report", The Royal Society, London (1983).

21. "The Tolerability of Risk from Nuclear Power Installations", Health and Safety Executive, HMSO, London (1988).

22. D R Grist, "Individual Risk - A Compilation of Recent British Data", Safety and Reliability Directorate, Report R125 (1978).

23. "Risk Criteria for Land-Use Planning in the Vicinity of Major Industrial Hazards", Health and Safety Executive, HMSO, London (1989).

24. M F Versteeg, "External Safety Policy in the Netherlands: An Approach to Risk Management", J of Haz. Matls., <u>17</u>, 215-222 (1988).

25. "Quantified Risk Assessment: It's Input to Decision Making", Health and Safety Executive, HMSO (1989).

Reduction of consequences

Kaspar Eigenmann

CIBA-GEIGY AG, CH-4002, Basel, Switzerland

Abstract - In the chemical industry the severity or consequences of a risk depend primarily on the properties and amounts of the chemicals used and stored and on the characteristics of the reactions and processes. Therefore, reduction of consequences must be a design criterion in the early stages of process development. Key factors are
- the synthesis route, which determines the chemicals and chemical reactions
- the process technology, which influences process conditions and the hold up of chemicals and energy
- the material flow concept, which determines the necessary storage capacity for raw materials, intermediates and final products.
Minimization of the risk potential in the course of process development is the most powerful tool for reducing the consequences because it eliminates the source.
Many technical safety measures aim also at reducing the consequences but without reducing or eliminating the cause or potential, e.g. in explosion protection: explosion venting, explosion supression or pressure resistant construction leave the potential for energy release unchanged but reduce with high reliability the consequences of an explosion. This type of consequence reduction often requires constant attention and maintenance in order to keep it effective.

Safety measures translate the findings of risk analysis into practical risk reduction. The ultimate goal is to reduce all risks of a chemical process or operation to an "acceptable" level.

In this context it is important to realize that the completeness and effectiveness of safety measures depends primarily on the quality of the risk analysis. No measures will be taken against risks not recognized in the risk analysis. Inefficient or excessive measures are the result of incomplete basic data or of incoherent or incomplete assessment of the risks with regard to severity and probability. Therefore a rather formalistic and systematic procedure for practical risk analysis is strongly recommended:

- The basic data - including those on reaction parameters - must be documented and the intended safe process conditions must be defined.

- A systematic method is required for the "search for hazards" to provide a complete list of potentially hazardous deviations from safe process conditions, which are evaluated with regard to probability of occurrence and severity of consequences.

- The resulting list of risks is the basis for the planning and realization of safety measures.

For risk reduction two principal routes are available: Reduction of possible consequences (or severity) and reduction of probability (or frequency). The two dimensions of risk are equivalent - if at all - only from an insurance point of view. In the real world however there is a principal difference to be recognized: Risk reduction by reducing possible consequences usually leads to an inherent and permanent improvement of the safety of a process or installation, whereas reduction of probability never goes to zero, therefore leaves a residual risk and often remains only effective as long as maintained and managed properly.

If - as an example - a solvent with a sufficiently high flashpoint is chosen, the danger of an explosion in case of leaks is eliminated. An often used alternative is to handle a solvent with a low flashpoint in an explosion-proof environment, which means that all sources of ignition should be eliminated by technical and organisational measures; thus the probability of an explosion is greatly reduced - but not to zero. It depends - among others - on the discipline with which the fire permit system is followed.

This example also illustrates that the reduction or elimination of consequence cannot always be the measure of choice. The solvent for a reaction usually can not be chosen freely; often only low-flashpoint-solvents are appropriate for a specific reaction. Therefore a combination of consequence and probability reduction is in general required to achieve the desired safety level of a process. The focus of this contribution is on consequence reduction.

In the chemical industry the severity of a risk is primarily a consequence of

- the properties of the chemicals, e.g. toxicity, volatility, flammability, reactivity

- the amounts of chemicals handled and stored

- and the characteristics of reactions and operations, e.g. exothermic potentials.

These factors can be influenced to reduce consequences principally

- by the selection of synthesis route, which determines directly the chemicals and reactions involved in the process

- by process technology, which influences the process conditions and the hold-up

- and by the material flow concept, which relates to the necessary storage capacity for raw materials, intermediates and final products.

A few examples shall illustrate these principal approaches.

To replace hazardous by less hazardous chemicals seems to be the most obvious strategy. And indeed, wherever we switch from a flammable solvent to an aqueous system, be it in a reaction or formulations of plant protection agents, we greatly reduce risk, not only with regard to the fire hazard, but also because of the higher heat capacity and enthalpy of vaporization of water, which are important barriers against exothermic runaway reactions.

In the chemical industry process technology is a very effective tool to reduce severity. Often a continuous process requires much less hold-up compared to batch processes. As an example, the diazotation of aromatic nitro-anilines, which involves a considerable decomposition risk, is traditionally carried out in batch reactors with a hold up of several cubic meters. The same reaction can be carried out in a continuous plug-flow reactor of 50-100 times less volume, yielding the same capacity. Thus the thermal decomposition potential is reduced by this factor.

If a hazardous intermediate can be produced at the rate it is consumed, storage of the intermediate may be greatly reduced. Phosgene is a good example. We have developed a process to generate phosgene from chlorine and carbon monoxide in a continuous reactor with variable output, which can be rapidly adjusted to demand. With this new technology, a storage tank of 15 tons of liquid phosgene could be completely eliminated. At present the only stored phosgen in this particular plant is the amount of gas in the pipe between the phosgene generator and the phosgene reactor.

Sometimes material flow or logistic considerations may help to reduce high hazard potentials. In one case we decided to move ethylene oxide reactions to the location of the supplier of ethylene oxide, thus eliminating storage and transport of this explosive and toxic chemical in a densely populated area. In another case, we recognized only after an accidental overfilling of a chlorosulfonic acid tank, that a 4 months supply was stored on site, although the supplier is quite close and very dependable. Adjusting storage to demand and to supply reliability not only reduces the hazard potential but also storage costs.

The type of measures discussed so far are of a rather fundamental nature. The choice of a synthesis route, even of the solvent, of the process-technology concept, the type of reactor - continuous versus batch - are all conceptual decisions which must be taken in an early stage of process development. For this reason, risk analysis must begin when process development begins, with the collection of basic data and the identification of the main principal hazards. We recommend to document these major risks in a hazard summary. Their reduction must be one among other important goals of process development. Hence risk analysis and process development are an iterative process. The hazard summary serves to keep these risks always in the foreground.

At this point a special comment on implications in the chemistry curriculum is necessary:

Collecting the basic data for a reaction, analyzing for the major hazards and elaboration of measures to reduce them should in the future be a standard procedure, which is taught at a certain level of the university curriculum and which from then on is regularly practiced in exercises - maybe even in the context of synthetic laboratory experiments. Only if we sharpen the chemistry student's awareness and skills in "risk reduction by means of process development" can we expect to change the general culture which too often tries to reduce risks, too late, when the process is introduced to the plant and neither chemistry nor the equipment can any more be changed.

Apart from reduction of hazard potentials by eliminating or reducing the amount of hazardous chemicals - or by replacing them - two other possible approaches shall be discussed:

- The hazard potential is distributed in smaller parts, separated from each other enough to prevent simultaneous activation ("distribution strategy").

- Or measures are taken to contain or reduce the effects ("containment strategy").

Fire compartments in warehouses are an example of the "distribution strategy", whereas second barriers of all kinds, from double walled pipes to fire water retention facilities demonstrate the "containment strategy".

Generation of phosgene at the rate it is used, is not always feasable or practical. In such cases cryogenic storage at or even better below the boiling point will reduce emissions in case of a leak quite substantially. This lowers the consequences of a phosgene leak considerably.

Dust explosion protection is a particularly interesting example. In this case 4 principal approaches are well known:

- Inerting is a measure which eliminates the hazard potential, no explosion can occur without oxygene.

- Pressure resistant construction eliminates the effects of an explosion. In this case, the hazard potential per se is still present and can be activated, but the construction prevents hazardous effects.

- Explosion venting directs the effect of the explosion to a safe place, where no harm is caused.

- Explosion supression detects an explosion early enough, then activates an explosion extinguishing system to supress the explosion before any damage occures.

Explosion supression depends obviously to a high degree on the reliability of the detection and the supression system. Therefore it represents an example where consequence and probability reduction must be combined.

In analogy fire detection and sprinkler systems are, if properly installed, measures to reduce severity or consequences. But they serve at the same time also to reduce the probability of a large fire, which demonstrates that certain measures may be looked at from a consequence, as well as a probability reduction point of view.

It must be emphasized that many situations exist where consequence reduction is not practically feasable, be it for technical, financial or principal reasons. In situations, such as air traffic, probability reduction plays a very important role. Sometimes the only other alternative would be to abandon the respective activity or project.

The benefits of risk reduction by reducing consequences are quite obvious:

- The risk reduction is of a principal nature, a step towards inherent safety.

- In general the measures taken are effective by themselves and do not depend on maintenance or special supervision.

- Processes and installations become less sensitive to human error or failure of technical safeguards (increased failure tolerance).

- Emergency measures (fire brigade etc.) must be planned with worst case scenarios in mind. Reduction of potential hazards may allow a shift of resources from mitigation to prevention.

- And last but not least in many cases an overall assessment of inherently safe processes and installations also shows an economic advantage.

On the other hand, consequence reduction is for several reasons a difficult task:

- It depends on an initial investment to fully understand the process and to analyze it systematically for major hazards in the very beginning of process development. This seems often to be difficult with the usual time pressure under which processes are to be developed.

- Safety measures to reduce concequences are often inherently related to the process itself. Safety can therefore no longer be a domaine only for safety specialists. It must be an integral part of the activities of chemists and engineers.

- The reduction of consequences is a multidisciplinary task. The integral approach must be supplemented with an effective integration of technical specialists into the project teams for process development.

- True reduction of consequnces often requires new, innovative or radical measures, which go beyond the traditional technologies and experiences. This requires an openness to new developments in the respective organization and persistent leadership from the part of the project team.

To overcome these difficulties is a challenge for the universities and technical academia to which the formation of our future leaders is entrusted, as well as for industry. Integration of safety (and environmental protection) into the general curriculum, development of the scientific aspects of safety and encouragement of "holistic thinking" could be contributions of the academia, where as industry needs to foster the adequate corporate culture, to set the right longterm oriented goals and to encourage innovation and development of safety technologies.

Safety measures
Minimizing probability

Presentation by Prof. R. PAPP
CORPORATE SAFETY AND ENVIRONMENT OFFICER ATOCHEM

Like many other human activities, the Chemical Industry has one imperative, namely to control the technological risks associated with its activities.

Because the products handled are often hazardous, because operations often involve high energy levels, risk control demands specific techniques, procedures and comprehensive rigour, the key to plant safety.

The Chemical Industry has made considerable efforts and devoted substantial financial resources to safety and environmental problems. It needs to go on doing all that is technically possible, economically reasonable and legally required. However, expressions such as "zero risk", "risk free" and "total safety" are practically meaningless. The idea of zero risk is utopian, and not only in the Chemical Industry.

Risk is normally defined by the potential hazard and the probability of the hazard's occuring. Clearly, the higher the potential hazard, the lower the event probability must be.

Many methods have been developed to estimate that probability. This presentation will be reviewing shortly some methods, as many has been already reviewed this morning.

These methods are supposed to be applicable to the various steps of the safety review. Their purpose is to incorporate all necessary measures into the design, construction, operation and maintenance of production units. Each step calls for different techniques.

A key point throughout is the exhaustive identification of potential hazards and their causes. Clearly, an unknown hazard will never be properly controlled.

Here again, several techniques have been developed over the past 15 years, including highly sophisticated procedures, laboratory techniques, computer simulations, etc. My presentation will give some examples of computer simulations, static or dynamic, which are an efficient tool to improve safety.

Hazard identification

Hazard identification is a continuous process during plant design. It starts at the research stage, with proper identification of the hazardous properties of each product, by products, and impurities liable to accumulate, properly identified, as for example polyperoxides with butadiene or vinyl chloride.

Major improvements have been achieved in the design of test equipment over the past 15 years, as shown this morning.

For example, continuous and fully-automated test reactors have been developed to identify explosive zones in mixtures, which are fed semi-continuously with incremented mixture ratios. The rise in explosive pressure is recorded by a computer which directly identifies the explosive zone and displays the explosive pressure ratio curves (figure 1) (Development made by the Decines process safety laboratory of Rhone-Poulenc France).

Microcalorimeters have been developed to improve our understanding of the kinetics of decomposition in self-decomposing products (figure 2 for hydrogen peroxide). The influence of decomposition catalysts such as metals can be assessed easily.

In 1979, the French "Union des Industries Chimiques" (chemical manufacturer's association) developed typical checklists enumerating all areas of concern to be considered when studying product-linked risks.

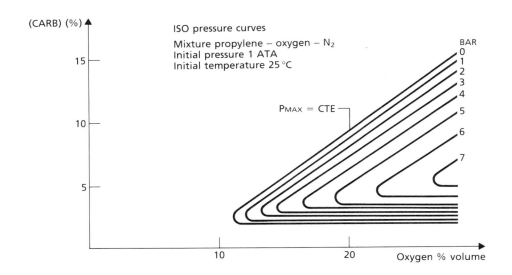

Source : Rhone-Poulenc Decines Safety Process Laboratory

FIGURE 1

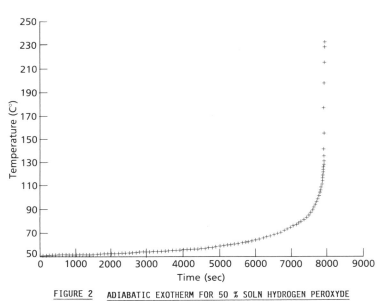

FIGURE 2   ADIABATIC EXOTHERM FOR 50 % SOLN HYDROGEN PEROXYDE
INFLUENCE OF CATALYSTS OF DECOMPOSITION

The same approach has been taken to reaction-linked risks.

Here again, a wide range of micro-reactors have been developed to determine :

> * Pseudo-kinetics of reactions.

> * Possible deviations of reactions.

> * Heat or reactions, pressure rise rate, etc.

The small scale of these reactors facilitates study of factors such as catalysts concentration, lack of refrigeration, accidental introduction of reagents or catalysts, etc...

All the major chemical companies have developed specific process safety research laboratories capable of using these techniques with maximum efficiency.

TABLE 1	TABLE 2
PREVENTION OF RUNAWAY	PREVENTION OF RUNAWAY
FOR PVC REACTORS	FOR PVC REACTORS
	CAUSE COMPENSATIONS DIAGRAMME
	FOR TEMPERATURE CONTROL FAILURE

TABLE 1

PREVENTION OF RUNAWAY

FOR PVC REACTORS

POSSIBLE CAUSES

EXCESS OF MONOMER
EXCESS OF CATALYSTS
ERROR IN CATALYSTS NATURE
REFRIGERATION SYSTEM FAILURE
AGITATOR FAILURE
TEMPERATURE CONTROL FAILURE
POWER SUPPLY FAILURE
HEAT INPUT DURING POLY PHASE

TABLE 2

PREVENTION OF RUNAWAY
FOR PVC REACTORS
CAUSE COMPENSATIONS DIAGRAMME
FOR TEMPERATURE CONTROL FAILURE

* CONTINUOUS PROCESS : AUTOMATIC SHUT DOWN

* TWO INDEPENDANT SYSTEMS FOR PRESSURE AND TEMPERATURE CONTROL WHICH REPRESENT SOME REDUNDANCY. OR

* REDUNDANT REACTOR TEMPERATURE CONTROL SYSTEM. OR

* TRIPLICATED SYSTEM 2/3 VOTING FOR TEMPERATURE.

* HIGH TEMPERATURE ALARM ACTING FOR MAXIMUM COOLING ACTION

* MONITORING THE REACTOR PRESSURE.

* HIGH PRESSURE ALARM.

* MONITORING THE COOLING WATER TEMPERATURE.

* SAME ACTIONS THAN COOLING FAILURE.

At the design stage

Hazard identification and definition of risk control measures is completed at the design stage by the process engineers, who may call for additional tests.

Again, recent development such as dynamic simulation of reactors and other items of equipment have spawned efficient techniques to reduce the probability of hazardous events.

Consider, for example, the risk of reaction runaway. Here, safety design will incorporate two steps :

* Identification of the causes of runaway and its consequences.

* Development of a computer simulation of the system by dynamic simulation, for example, by using the SPEEDUP* code, developed by Professor SARGENT, Imperial College, LONDON.

* Evaluation of risk control devices and work to improve their efficiency.

Tables 1 and 2 illustrate the diagramme causes -process safety measures- capable of being considered for a particular case, in these tables the polymerization of PVC.

(*) SPEED UP is able to handle 8000 differential equations. There is therefore no mathematical limits but only limits in the basic physical or chemical properties or kinetics to enter into the code.

Figure 3 shows the pressure temperature curves of runaway obtained from the simulation.

The simulation computer code enables us to check all the process safety measures and their efficiency.

Figure 4 shows for example the impact of an additional water refrigeration tank in the refrigeration loop in case of power failure for a Ionic polymerization runaway. In this case, the hold up of frigories is reducing considerably the pressure rise slope. Therefore additional time is available to implement other safety measures.

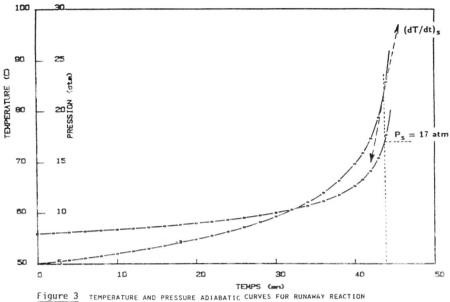

Figure 3   TEMPERATURE AND PRESSURE ADIABATIC CURVES FOR RUNAWAY REACTION

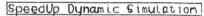

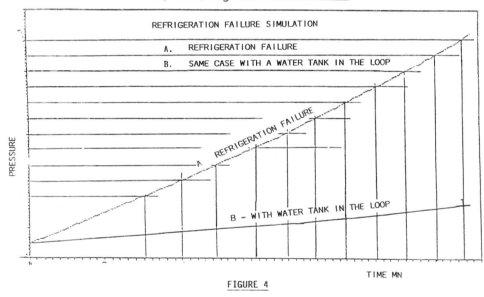

FIGURE 4

PRESSURE ADIABATIC CURVES FOR IONIC POLYMERISATION

The prevention of runaway for this polymerization reactor is based on six items :

* Three different inhibitor systems.

* Emergency power supply.

* Additional refrigeration tank and pump to lower the pressure rises slope as shown in figure 4.

* Rapid degassing system to the flare through a blow down tank.

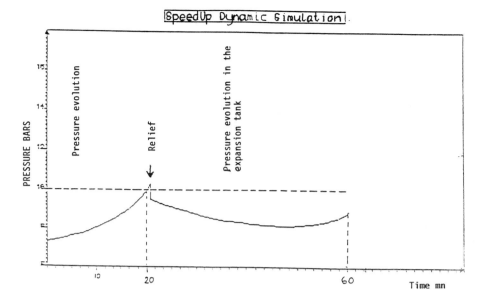

FIGURE 5   RUNAWAY SIMULATION WITH CONFINEMENT OF THE RELIEF EXHAUST IN AN EXPANSION TANK

Figure 5 shows the simulation of a runaway with recovering in a blow down tank.

At the construction stage

Plant safety depends primarily on proper specifications and procedures, and on rigorous control of their application.

The wrong metal in certain items of equipment, cracks in the heated welding zone, failure to check them properly, failure of a crane, or mobile trucks or cranes crashing onto equipment, etc. undoubtedly increase the risk of accident.

Suitable procedures and follow-up are essential to ensure that the plant is built as designed, according to the specifications and drawings. Particular care is needed when construction is in progress in or near an operating plant.

HAZOP* Safety Reviews must be carried out before approval of all documents, and especially piping and instrument diagrams, additional to the earlier process safety studies, described as "preliminary risk analysis" by the French U.I.C in its publications.

While in principle the HAZOP study is able to identify and review all relevant situations, timing too is a key factor : Basic Safety Concepts such as the continuous feed of a reactant in order to prevent runaway, must be taken into consideration at an early stage of the design process, even earlier at the research stage.

With a good preliminary risk analysis, the HAZOP review will consider the possible combinations of failures, which can only be identified on a P and I diagram, but will limit the changes to be made to a minimum.

Also considered at this stage are possible human failures and the relevance of the safety instructions.

A host of codes, standards, norms and guidelines are applied during the construction stage. All these documents provide considerable help to engineers. However, codes, standards and norms often hinder technicians from getting on with more practical, more detailed studies which may be required for sound, safe design.

Another important point to consider in the design and construction stages is the question of lessons to be learned from past experience.

Innovations always entail some additional risks and must be incorporated only if necessary, and with particular care.

This policy has been applied consistently for many decades by chlorine manufacturers, resulting in a reasonably good safety record.

Operation and maintenance

Many catastrophes are caused by human failure, chiefly by non-compliance with procedures and safety instructions.

Safety starts in the minds of the operatives and maintenance workers who have do deal with hazardous products and processes.

(*) Hazard Operability Study

The requisite degree of rigour will be applied only by people who understand the need for it.

Obviously though a satisfactory degree of safety can be achieved only through a broad set of internal safety rules, familiar to all and subjected to periodical audits.

Procedures such as :

* Modification procedures

* Intervention procedures

* Entry permits

* Inspection procedures for vital plant safety equipment

* Safety audits

etc.

are important and must be applied in all hazardous plants.

Another important document, a Company Policy Guide, signed by the Chairman himself, will help to emphasize the priority that the company attaches to these issues.

What probability ?

All of the measures briefly reviewed above are designed to prevent hazardous reactions and accidents from happening.

Having identified the key hazardous events in a plant, it is next necessary to check that the probability of their occurence is sufficiently low in regard to their consequences.

An event can be quantified by drawing a fault or an event tree for example. All primary causes will be identified until they can be linked to a probability (Figure 6). The significance of the degree of probability is indicative only but valuable for the design of safety measures to assess the degree of safety.

The French Association of Chemical Manufacturers has defined six levels of seriousness and six levels of probability permitting them to be semi-quantified (Tables 3 and 4). This procedure is precise enough to assess the degree of safety in regards of the consequences of an event.

Another way to assess the probability of an event is by studying past experience.

There have been 32 "BLEVE"* of LPG pressure storage tanks in the world in the past 30 years, but only two in Europe. We may therefore calculate a probability of 10^{-5}/year for this kind of event in light of past experience worldwide, and 10^{-6}/year in Europe.

But the LPG storage tanks in consideration were of the old design, dating back 30 years. A new design for LPG storage tanks under pressure, incorporating all safety measure provisions such as foam in the bund beside the vessel and not underneath fireproof insulation, etc. will substantially reduce the probability of BLEVE.

The Health and Safety Executive calculated the probability according to a fault tree and estimated the BLEVE event probability of a LPG storage tank to be less than 10^{-7}/year[1]

EXAMPLE OF SEMI QUANTIFIED EVENT TREE

FIGURE 6

EVENT : RELIEF VALVE BLOWING ON A DISTILLATION COLUMN

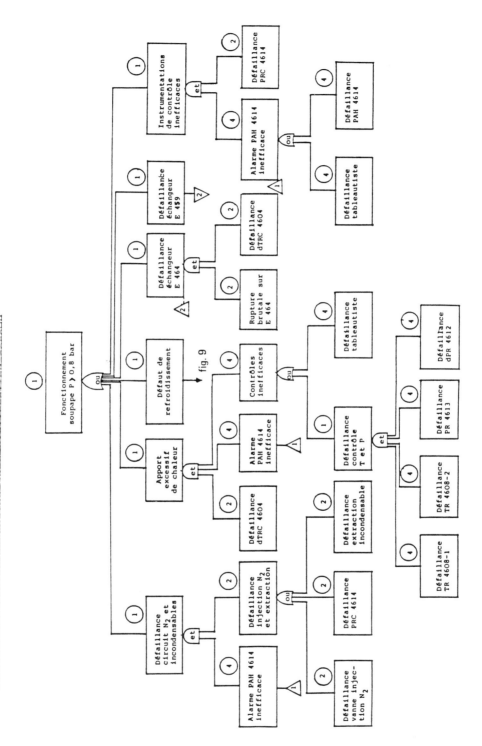

T A B L E 3	T A B L E 4
CLASSIFICATION OF EVENT PROBABILITIES IN SEMI-QUANTIFICATION (UIC)	**CLASSIFICATION OF EVENT GRAVITY IN SEMI-QUANTIFICATION (UIC)**
LEVEL 1 EXTREMELY RARE	LEVEL 1 THE EVENT INVOLVES NO INJURIES AND NO DAMAGE.
LEVEL 2 VERY RARE	LEVEL 2 THE EVENT INVOLVES NO INJURIES AND NO DAMAGE TO THE EQUIPMENT BECAUSE CORRECTIVE MEASURES ARE SIMPLE.
LEVEL 3 RARE	
LEVEL 4 POSSIBLE	LEVEL 3 THE EVENT NECESSITATES QUICK ACTION TO AVOID INJURIES AND DAMAGE.
LEVEL 5 FREQUENT	LEVEL 4 THE EVENT CAN LEAD TO SERIOUS CONSEQUENCES LIMITED TO THE UNIT ITSELF.
LEVEL 6 VERY FREQUENT	
	LEVEL 5 THE EVENT CAN LEAD TO SERIOUS CONSEQUENCES LIMITED INSIDE THE PLANT.
	LEVEL 6 THE EVENT CAN LEAD TO SERIOUS CONSEQUENCES OUTSIDE THE PLANT.

With suitable fireproof insulation, the risk will be reduced still further and becomes very small.

This analysis may be compared with the risks inherent in an underground LPG storage tank, where we have no past experience to go on, and where other risk such as ground subsidence, external corrosion, lack of inspection, etc. need to be taken into consideration. It may be true that the risk of a BLEVE in such a storage tank will be lower, but the reduced risk of accident in general is questionable.

Quantification -or at least semi-quantification- is necessary in order to determine the low probability of a major risk.

But in all cases, the value of any such probability for a given equipment and plant design will always depend on the reliability and quality of risk management at the site itself. Technical design may give this probability. Risk management only will guarantee the level of risk. This is clearly demonstrated by analysis of major accidents occurring in recent years.

To conclude, the way to reducing the probability of major hazards is by framing and implementing a thorough policy inside the company, involving not only safety specialists, obviously, but also the entire personnel concerned, from research through plant maintenance inclusive.

A large number of methodes, well adapted to each step of the development of a new production unit are available for the technicians, and provide a considerable help to ensure safety

* BOILING LIQUID EXPANDING VAPOR EXPLOSION
(1) HEALTH AND SAFETY EXECUTIVE PUBLICATION SRD/HSE/R 492

Safety management
A view from industry

Ir P de Voogd

Shell Internationale Chemie Maatschappij BV, The Hague, The Netherlands

INTRODUCTION

The damage resulting from a chemical runaway reaction can be considerable as is shown in plate 1.

Nobody wants accidents to happen. So how can we avoid them?

As my contribution to the workshop I will describe the way my company deals with risk reduction in practice and the systems it uses in general to achieve high standards in safety, health and environmental protection.

I work for a Shell company in the central offices in the Hague. My job is adviser in Health, Safety and Environmental matters for Chemicals manufacturing. I provide this service to the Operating Companies of the Royal Dutch/Shell Group of companies worldwide and to the project department in the centre. As such I get involved with a great variety of chemical installations ranging from small plants to large complexes; there are some 120 Shell plants and the number is still growing. Shell companies operate a multitude of processes in a variety of countries with large differences in culture.

For each of these, safety management must be adapted to the local circumstances in order to achieve the best results. The basic framework however, always applies.

fig. 1

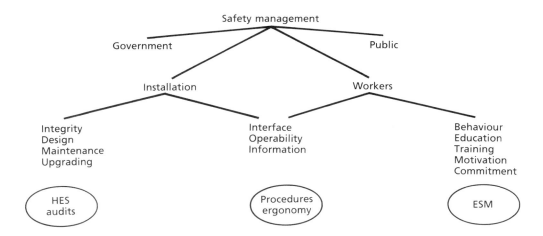

THE SAFETY MANAGEMENT SYSTEM

Management of safety involves installations and workers - both company and contractor personnel. It also involves cooperation with the government bodies and good contacts with the neighbouring communities (fig. 1).

I will discuss the management of installation integrity, of people and their behaviour and of the interface between them because these are the three building blocks to achieve good safety.

Shell companies are expected to adhere to HSE policies with the Shell Business Principles and to follow the Group guidelines. For affiliates and joint ventures we want assurance that a system which provides equally high standards is in place.

The Shell Service Companies in The Hague (SICM/SIPM) develop and issue guidelines on the various HSE subjects for use by all Group Companies. The system to ensure that the guidelines are followed consists of procedures for project development and approval, regular audits and yearly appraisals of performance.

All Group Companies and their facilities are regularly audited. This is a service which the Service companies provide worldwide. The audits check on the actual situation in aspects of management, procedures, awareness, technical status etc. Recommendations for improvement are presented when considered necessary. The result is a continuous improvement in HSE standards and performance.

Let's look at the three subjects in some detail.

INSTALLATION INTEGRITY

To ensure that a plant is safe when it is constructed or acquired, and that it will remain safe during its total life, it is necessary to:

> develop a good project
> maintain the technical integrity
> continuously seek to improve

- develop a good project:

Fundamental for technical integrity and safety is to design and construct the plant in such a way that it can cope with all the foreseen conditions to contain the chemical substances we want to process. Much will depend on the choice of process and on the properties of the chemicals used and produced in the process.

Examples:

1)   Structures for the plant may have to cope with a soft subsoil or with earthquakes. Civil engineering will have to make sure that the structures protect the equipment from damages (Moerdijk underground pipe leakages).

2)   Disasters like the LPG fire at Feyzin have caused the industry to review the design and layout for LPG type storage to eliminate flame impingement.

Process safety and safeguarding needs particular attention in chemicals manufacturing because of a variety of hazards such as product flammability, toxicity and the possibility of strong exotherms.

The safeguarding philosophy followed by Shell companies is that safety systems should be independent from process control systems (redundant) so with modern electronic systems, the solution is either to have the safeguarding in conventional (hardwired) design or to install redundant fixed program computers which offer reliability for fail safe action with power dips etc.

To deal with health, environmental and safety requirements in Chemicals manufacturing project preparation, we use a studyteam approach (HES) in which specialists support process engineers and designers (fig. 2). The reason for a team approach is to optimise communications between the different professions. The HES studies should use methods and procedures which are sufficiently flexible to allow good studies for non-routine projects. One main objective of these HES studies is to minimise risks and cater for the contingency of the effects of residual risks.

fig. 2

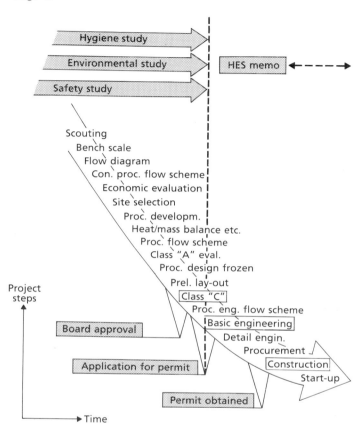

For a good project we need to know the environmental conditions, the infrastructure of the location or country, the availability of qualified personnel, the supply position of spare parts, etc. The quality of the work that goes into the preparation depends on people in the project team; the safety on what the management wants.

If sophisticated equipment or skilled technicians are not available in the country where the plant is to be built, this will require the project to use simple equipment that can be repaired or replaced locally. The plant can also be just as safe.

All projects and modifications whether developed in the centre or locally are subject to check on HSE acceptability before management approval is given. The HES studies will of course greatly assist in such a check.

- **maintain the technical integrity**:

A management system should be put in place to ensure that proper inspection and maintainance takes place on all parts of the installation, that safeguarding systems are reliable and available at all times during the operation, and that any modification is thoroughly scrutinised for acceptability.

Maintaining technical safety and integrity is dependant on the way the installation is operated and maintained during its life.

Preventive maintainance, methods and rules for inspection and action on findings, interface operations and engineering work, testing of safeguarding systems, start-up and shut-down procedures are all relevant aspects.

- **continuously seek to improve**:

We improve the technical safety when new knowledge is available for example from research and development or following lessons from accident investigation. This knowledge is not only incorporated in new projects but also used for upgrading existing installations. We have a central system to collate and analyse the data and to disseminate it to all who need to know.

example:

> The investigation of a fire in an ethylene oxide plant due to soaked insulation lagging reacting exothermally has led to a number of recommendations eg one for replacement of lagging material for that purpose.

The skill and knowledge of the people carrying out these tasks should be at a high level. We have a tradition of job rotation between Operating Companies and Central Offices that brings about mutual understanding and a rich experience in the various jobs viz. the researchers, the process and project developers, the design engineers, the constructors, the plant managers, the maintainance engineers (fig. 3).

fig. 3

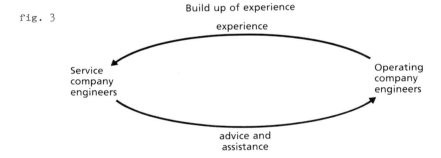

INTERFACES

Traditionally, the interface between worker and installation has been covered by instructions and procedures. Nowadays we do much more.

This is a subject for attention from both the designer and the plant manager. The designer is to provide a plant which is easy to operate and provides the necessary safety. The plant manager is responsible that the personnel has the right skills and that the right procedures are in place.

The operability and safety of a plant is strongly influenced by the layout of equipment and instruments, by the information presented to the operator and by the possibility for the operator to control the process and its deviations. He should have the feeling that he can master all situations.

We use ergonomics in design of control rooms, in process control and information systems, in positioning of equipment and in plant layout. For the latter the use of small scale models of the plant is of great value. We are still learning about ergonomics, but it has already brought important benefits.

With respect to failures, studies have shown that although well trained operators are perfectly capable to handle normal situations ie steady operations, start-up and shut-down, familiar or recognised upsets, they still make errors. The frequency of these errors is stongly influenced by the working environment: human relations, discipline and quality requirements, technical conditions (fig. 4).

fig. 4

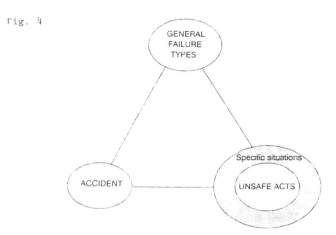

Reduction of this frequency will result in reducing the probability of accidents. I will come back on how to do this.

The operator response becomes quite different and in fact unreliable when unknown situations occur such as unprecedented upsets in the process. This apparently is because the human brain does not usually use logic for fault diagnosis under time pressure (ref Prof Wagenaar). To some extent improvements can be achieved by using support systems to help identify the cause of a deviation or fault.

We train with the aim to reduce the probability of failure. Part of this training consists of exercises with scenarios of the events which can be envisaged and also tries to prepare the operator to deal with the unexpected.

The consequences are reduced by a number of design measures such as minimum inventory of hazardous chemicals and devices to contain or minimise releases eg emergency shut-down controls, remotely operated valves to block in sections of the plant, water curtains to avoid dispersion.

PEOPLE BEHAVIOUR

We have found that all industrial accidents have an element of human failure, with the majority it is even one of the prime causes. Nevertheless, in Quantitative Risk Assessment, this failure element is normally not taken into account because it cannot be quantified.

To achieve a safe operation everybody in a company should be convinced that safety is important and that they can personally contribute. This conviction needs to start at the top and management should clearly demonstrate that it is committed to a safety programme. This commitment will gradually permeate through the whole organisation.

Shell companies select and train personnel thoroughly in job content and in safety (HSE) aspects. Motivation to ensure high safety standards is promoted throughout the organisation with the aim to avoid personal injuries and plant accidents. This applies to all staff: the managers as well as the workers.

That safety is important, does not need much explanation. It suffices to point to the consequences of accidents in terms of suffering of people, damages to property or nature and to the company's reputation. The cost of accidents is always high.

Shell policy has always been to maintain high safety standards but the progress in performance having been unsatisfactory in the 1970's (fig. 6), we decided to launch a new programme in the early 1980's which we called 'Enhanced Safety Management'. We contracted the help of external experts in particular for the human behaviour aspects.

For an organisation like the Shell Group with some 120,000 employees working in a great variety of member Companies in so many countries, it is quite a task to reach everybody but we are convinced it is worthwhile.

We are devoting a great deal of management attention and effort to get people committed to safety since it has become clear that in every accident there is an element of human failure and in most it is one of the predominant causes. Whether it is a single personal injury or a major calamity, every accident investigation reconfirms this

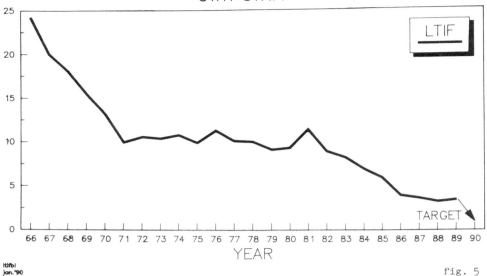

fig. 5

finding. The failure may be due to lack of skill or knowledge, or it may be related to a deviation from the right working method because of habit or belief it will not cause harm (acceptance of risk), "it always went right".

Changing this behaviour is not easy. There are four strategies based on psychology (ref Prof Burkardt) to motivate safe behaviour and to demotivate the unsafe one (fig.6). It requires conviction, discipline and patience. There are various management methods to achieve this and the most successful combination will depend on the local culture and on the progress made with enhancing safety. A very powerful method aimed at eliminating unsafe acts is to organise for the line managers to visit the workplace regularly, observe people at work and discuss the directly related safety aspects with them.

At all times the responsibility for safety should be with line management; they have the power of decision. Safety advisers have a supporting role. They should of course be knowledgeable and competent as their advice can be of crucial importance. In the end everybody is responsible for safety. A company can only work safely if every individual fulfils his task in a responsible way and looks after his own safety. Management alone cannot ensure 100% coverage.

STRATEGY I	STRATEGY II
INCREASE POSITIVE REINFORCEMENT FOR SAFE BEHAVIOUR	REDUCE NEGATIVE REINFORCEMENTS FOR SAFE BEHAVIOUR
STRATEGY III	STRATEGY IV
ACTIVATE NEGATIVE REINFORCEMENTS TO UNSAFE BEHAVIOUR	ELIMINATE POSITIVE REINFORCEMENTS FOR UNSAFE BEHAVIOUR

fig. 6

Getting people to "think and act safety first" is therefore the objective which is expected to achieve the best improvement in safety performance. Education and training of personnel together with experience on the job will provide them with the required knowledge and skill. The safety aspects form part and parcel of the training. Motivation for safety can furthermore be raised by a system of appraisal and rewards which favours the safe performance over the "quick and dirty" approach.

Experienced personnel trained in this way becomes a very valuable asset in the company.

Coming back to the safety management structure, we have seen that as we have to work through people, the safety level we achieve will depend on the commitment and capability of everyone in the organisation to work towards safety. Achieving that is an essential part of safety management.

REFERENCES

1. A J M Groenewegen What happened? Diagnosing unfamiliar real life situations. Thesis Univerity of Leiden, 1990.

2. W A Wagenaar & J Groeneweg Accidents at sea: multiple causes and impossible consequences. International Journal of man-machine studies, 1987, 27.

3. W A Wagenaar The cause of impossible accidents. University of Amsterdam, 1986.

4. F Burkardt How to encourage safe behaviour. Frankfurt, 1985.

5. W A Wagenaar Human failure. Leiden, 1983.

Plate 1 The damage from a chemical runaway reaction.

Stewardship of chemical production risks

William W. Lowrance

Life Sciences and Public Policy Program, The Rockefeller University, 1230 York Avenue, New York, NY 10021, USA

INTRODUCTION

At the outset we should recognize two things as starting points. First, chemical facilities inherently tend not to be easy to love; most of them, after all, are rather smelly and unsightly, hazardous to some degree, involve massive and intrusive transport through communities, are mysterious, and, for most people, recall their least favorably remembered school science courses. Because of its history of accidents -- and its poor handling of some dramatic and serious accidents -- and less than full candor, the chemical industry is not entirely trusted. And because of the specialization in modern society, reasons for having to produce and process chemicals aren't very familiar to the public. Any discussion of the issues has to acknowledge this negative public attitudinal background, regardless of whether it is deserved.

Second, society's management of this complex of issues involves not just science and technology, but also profound ethical, economic, and other value-laden considerations. Engineers and other technical specialists must, to be sure, select proper processing conditions and design optimal valve systems and conduct risk assessments. But also they must work with other managers and with the public and its representatives and authorities to decide what degree of precaution is warranted, how much of what risks are acceptable, and which procedures to follow in informing and reassuring the public and working to minimize risks.

In addressing these controversial topics, I feel obliged to reveal my own views at the beginning:

- No activity in human life can be risk-free; indeed, risktaking for benefit is the essence of human striving.

- Chemical production and processing are absolutely essential in modern life. We must try very hard to minimize the risks in chemical industry. But the only way to reduce any industrial risk *absolutely to zero* is to abandon the whole process or product. We must expect to have to live with some industrial risks and their incremental addition to our overall risks.

- We must guard against merely transmuting risks from one form into another.

My task in this paper is to survey some broad aspects of "risk" in general, and then suggest how they can be applied to sizing-up and managing the risks in chemical production.

A FEW "RISK" BASICS

Risk is a compound estimate of likelihood and severity of harm. No matter what the specifics, a key to dealing with risks is recognizing three separable modes of analysis and action:

Risk assessment, *description* of the likelihood and severity of threat;

Risk appraisal, *evaluation* of the personal or societal burden from the risk, the costs required for protection, and the payback expected from risk-reduction investment; and

Risk response, *prescription* of "what to do about" the risk.

These modes are distinguishable but interrelated. Risk assessment finds scientific facts; risk appraisal weighs consequences of the facts in light of personal and social values; and risk response makes action-decisions based on the facts, values, and pragmatics.

RISK PORTFOLIO ELEMENTS

On major personal or social problems we build up a composite set of descriptions, evaluations, options, preferences, and so on, which can be thought of as a portfolio. This composite then is examined and debated by concerned parties. The following elements have to be assembled to constitute the portfolio on any risk. They should be readdressed and revised as progress occurs in factual understanding, social concern, or possibilities of remedy.

Risk assessment ("How much harm is incurred or is expected?")

- Characterization of the nature of the hazard;
- Effects on individuals from exposure;
- Individuals' exposure;
- Inventory of the population(s) exposed;
- Relationship between degree of exposure and degree of effect (dose--response);
- Prevalence, severity, and distribution of the risk in exposed populations and in society as a whole.

Risk appraisal ("How much to be concerned?")

- Personal and societal burdens from the risk;
- Benefits and detriments associated with the source of hazard;
- Comparison against similar risks and against risks from alternatives;
- Costs and paybacks expected from protective investments;
- Ethical consequences of actions (or inaction).

Risk response ("What to do about it?")

- Practicalities in modifying the risks (technical feasibility and effectiveness, costs, political constraints, timing);
- Overall comparative promise of the various response options;
- General climate of public opinion.

Appraisal is the element that bridges between scientific factfinding and social response. It estimates the risk burden in social context, then evaluates the payback from current or contemplated protective investments. Risk appraisal is one of the most underattended aspects of the chemical production risk portfolio.

It is important to develop the elements of such a portfolio, constantly strive to improve the quality of the least-well-understood or least-fully-agreed-upon elements, and discuss the portfolio publicly. This helps build accountability of those in charge, avoid surprises, and increase the apparent legitimacy of the enterprise. Summarizing the portfolio explicitly and on public record gives anyone interested the opportunity to endorse or criticize aspects of the way things are being managed.

DEFINING BOUNDARIES

The outcomes of all analyses are very sensitive to the boundaries that are adopted, such as which issues are considered, and the analytic assumptions that are incorporated. Debate has to begin by asking: What are the dimensions of The Risk Problem? For chemical enterprises, typical bounding issues include:

- Operations limits: Include transporting of chemicals? Storage? Recycling, disposal? Anticipate routine leaks, spills, fires?

- Misuse or abuse causes: Anticipate malfunctioning of processes? Worker error? Sabotage?

- Associated physical causes: Anticipate nearby fire, firefighting? Railway or docking accidents?

- Associated natural catastrophic causes: Anticipate flood, earthquake, mudslide, high winds, etc.?

- Routine fugitive emissions: Take account of routine trace releases (such as evaporating solvents)?

- Other wastes: Be concerned about ultimate emissions (such as incinerator emissions, or river heating) and residues?

- <u>Catastrophic chemical releases and explosions</u>: Consider worst accident possibilities? Worst-worst...? Integrate consideration of moderate-probability--moderate-severity risks with low-probability--high-consequence risks?

- <u>Subjects at risk</u>: Be concerned about workers? Consumers (including industrial consumers)? Plant neighbors? General populace?

- <u>Environments at risk</u>: Which ecosystems and biota to consider? Include distant environments?

- <u>Derivative effects</u>: Consider secondary environmental and health consequences? Tertiary?

- <u>Societal structures at risk</u>: Worry about employment? Disruption of the social fabric? Property-value structure? Archeological or patrimonial inheritance?

- <u>Time horizons</u>: How far into the future to look, for what effects?

In making optimal protective decisions these variables may sometimes be traded-off against each other in an effort to alter aspects of the risk -- such as minimizing the worst-catastrophic risk, or reducing risk to the highest-exposed population, such as workers. Or high risks in transport might be reduced, at the expense of having to control against increased risks in on-site recycling. By pursuing these questions one may become aware of risk displacements -- changes wherein one risk is reduced, but by being transmuted into another risk in another domain.

The establishing of these boundaries and assumptions -- which gives guidance to engineers and economists who perform the analyses, and which sets the ground for social debate over the specifics -- clearly is an ethical and political matter. One of the principal values of regulatory and public review of analyses is debating these defining issues.

GENERAL ATTITUDES

In order to generate perspective on chemical processing risks, it is important to try to discern some perceptual themes. Though conjectural on my part, the following generalizations about the public's attitudes and values have wide currency and help explain many actions.

- Fundamentally we conceive of ourselves as risk-buffering societies: communally taking precautions, and arranging remedy for and compensation of those who suffer unduly.

- We strongly prefer to choose our own risks, and resent having risks involuntarily imposed on us. We try to extend respect for voluntariness to everyone, via the ethical principle of consent.

- We are more willing to allow others to undertake risky actions if the potential health and cost consequences are confined to the risktakers.

- We are more willing to condone a risk-imposing activity if direct benefit accrues to those who bear the risk, or if those who bear the risk agree to accept monetary or other compensation from those who benefit, or if the risk is a side-effect of an activity that delivers widespread common benefit to the community.

- Emotionally and politically we find it easier to endure risks that fall on victims unspecifiable in advance or unknown to us -- "statistical risks" -- than those that menace identifiable victims or people we know personally.

- Immediate and obvious threats close to home tend to attract more vigorous citizen response than those far away, less certain, or latent.

- Catastrophic, lumped losses are emotionally harder to endure than a series of scattered smaller losses, even if the smaller losses are equivalent in sum. For policy, the issue is whether to "overprotect" against disasters because they are so disruptive and disturbing.

- Because as individuals our expertise and power are limited, we are as concerned that risk issues be supervised by competent, frank, honest, responsive, trustworthy leaders and authorities, as we are concerned about particular risk outcomes.

Viewing chemical production activities in light of these themes, we observe them as tending to be somewhat diffuse and having important communal implications; imposing rather involuntary risks (even for plant employees); delivering clear local economic benefits but levying local burdens, while providing benefits for the larger society; routinely bringing ordinary risks, and perhaps holding catastrophic potential; and being technically complex and under the control of experts and specialized managers.

THE SOCIAL RISKBEARING CONTEXT

Chemical enterprises have to be examined in very broad perspective, having to do especially with relative risk, assurance of precaution, and trustworthiness of those managing the activity. The important generic questions are:

- Generally, is the production activity worthwhile and responsible?

- Has every reasonably-expectable precaution been taken to reduce the risks and cope with harms if they are incurred?

- Are the risks similar to other, familiar and routinely-borne risks, or are they strange or exceptionally high?

- Are the people potentially at risk informed of the risks, and of how to take personal precautions? And have the people most at risk consented to bear the risks?

- Do those who bear the risks, benefit from the activity that generates the risks (or willingly accept compensation from those who benefit)?

- Are the authorities in charge of managing the activity and controlling its risks competent, honest, well-intentioned, and trustworthy?

- Is demonstrable progress being made in managing the risks better?

Many of these can be answered in commonsense language. Debates may center around "worthwhileness," "reasonably expectable precaution," and informed consent.

All of these matters are of concern not only to directly-affected people, such as plant employees, and to "the public" and its representatives and surrogates, but also to plant managers and government officials who must exercise civic duty. Concerned citizens may well be seeking, not detailed debate on specific questions, but general reassurance.

Social justification of chemical production must center on reducing the risks to very low levels *so as to accrue the societal benefits*. All risk-reduction efforts should be viewed as societal investments.

A major difficulty for chemical enterprises is that because most members of the public lack firsthand experience with primary production, they fail to appreciate the importance of producing the commodity, intermediate, and fine chemicals that are required for securing the agricultural, transportation, and other benefits that they are avid consumers of. Producers don't produce if consumers don't consume. One doesn't have to "buy" all of the essayist Wendell Berry's message to agree with his admonition that, "We cannot continue to divide the world falsely between guilty producers and innocent consumers." Every possible effort should be made to remedy this unawareness.

STRATEGIC RISK REDUCTION

My "sermon" -- the approach I advocate for dealing with any risks to health -- can be summarized as follows. Whenever confronting any risks, we must:

- Be *strategic*, striving to reduce worst threats first (and debate what "worst" means in given instances);
- View risk responses as social investments, and favor programs that promise the highest humanitarian return-on-investment.

This requires taking *comparative* approaches:

- In assessing risks;
- In appraising risks;
- In responding to risks;
- In consulting with the public.

The strategic implications must be reflected in:

- The macro societal (regulatory, etc.) agenda and priorities;
- The management of specific risk-response programs;
- Planning of risk research and education.

And the dynamics must involve:

- Managing uncertainties explicitly;
- Scaling all response actions to the societal and personal stakes;
- Developing plans and programs in consultation with the affected publics;
- Addressing ethical complexities.

What this means for chemical enterprises is conducting systematic risk appraisals, setting priorities carefully and openly, and implementing these priorities explicitly in engineering design programs, employee training, day-to-day management practices, negotiations with government authorities, and working with surrounding communities to manage the facilities and prepare for emergencies.

RECENT CONSTRUCTIVE INITIATIVES

In the past few years many initiatives have been undertaken in many countries to reduce health and environmental risks from chemical production. The following are several whose examples and "lessons" are being widely adapted.

Item: National industry programs implemented at local level. More than a thousand American corporations have implemented commendable Community Awareness and Emergency Response (CAER) programs, developed by the U.S. and Canadian Chemical Manufacturers Associations, to sensitize and inform plant neighbors, hear their concerns, and improve emergency preparedness. Several hundred U.S. corporations and companies in other countries recently have adopted Responsible Care initiatives, which include guiding principles, codes of management practice, self-evaluation, public advisory scrutiny, and executive leadership improvement. All of these programs have been conducted in concert with government programs, and they involve public consultation. Responsible Care programs are outlined and coordinated nationally but implemented locally, tailored to the situation.

Item: A national professional engineers' program. The American Institute of Chemical Engineers (the professional society) established a Center for Chemical Process Safety to develop and evaluate technical and operational practices to predict, prevent, and mitigate major accidental releases of toxic chemicals. It has conducted some technical studies of vapor release and dispersion, held conferences, and published thorough book-length guidelines, *Chemical Process Quantitative Risk Analysis* and *Technical Management of Chemical Process Safety*.

Item: A chemical-regional program. A National Institute for Chemical Studies was formed by a consortium of the industries, citizens, and government bodies in the Kanawha Valley of West Virginia, to manage and reduce the risks to the people who live within that major chemical corridor. The Institute, which despite its broad name mostly acts locally, has sponsored projects to measure air toxics, conduct health surveys, catalyze voluntary reduction of toxic emissions, improve emergency preparedness, develop criteria for deciding between evacuation and sheltering during accidents, and conduct a variety of community awareness and informal dialogue programs.

Item: International programs. The United Nations Environment Programme (UNEP) provides a variety of technical information and other services, especially to developing countries, as has the International Environment Bureau. UNEP has disseminated the CAER precepts to many countries via an Awareness and Preparedness for Emergencies at the Local Level program.

Many analogous initiatives have been undertaken. A variety of European and international accords have been established for managing these problems. Under "right-to-know" guarantees, programs are underway to provide much more information to the public. Risks in chemical transport are, tardily, being addressed. Certainly this IUPAC Workshop is a contribution.

ELEMENTS OF RESPONSIBLE CHEMICAL PROCESS MANAGEMENT

A great many techniques and procedures were ably discussed in detail in the symposium, and there is no reason to repeat them here. In synopsis, the essential elements of responsible corporate chemical process management are:

- Agreed-upon risk-control goals and objectives;

- Clear chains of responsibility and accountability;

- High general technical competence;

- State-of-art approaches to assessing, appraising, and responding to risks;

- Robust decisionmaking procedures;

- Adherence to strong codes of corporate conduct and guidelines of professional practice;

- Conformance with both the spirit and letter of regulatory and other legal stipulations;

- Ongoing personal sensitization, training, and experience review;

- Protection, preparedness, and performance audits;

- Post-accident and -incident investigations, and improvement via feedback;

- Attendance to infrastructural needs (such as technical databases, experience-sharing networks, risk-reduction techniques);

- Constructive working relations with the newsmedia; and

- Active dialogue with workers, special-interest groups, and the general public and its leaders.

A MODEL SET OF PRINCIPLES

Codes of conduct can help articulate these matters. In this regard the Chemical Manufacturers Association (CMA) "Guiding Principles for Responsible Care" [which I have not been involved in developing, but admire] deserve being quoted in full. They are ideals, and surely their implementation into everyday practice must not be easy. Nevertheless -- commendably -- CMA members pledge:

1. To recognize and respond to community concerns about chemicals and our operations.

2. To develop and produce chemicals that can be manufactured, transported, used and disposed of safely.

3. To make health, safety and environmental considerations a priority in our planning for all existing and new products and processes.

4. To report promptly to officials, employees, and customers and the public, information on chemical-related health or environmental hazards and to recommend protective measures.

5. To counsel customers on the safe use, transportation and disposal of chemical products.

6. To operate our plants and facilities in a manner that protects the environment and the health and safety of our employees and the public.

7. To extend knowledge by conducting or supporting research on the health, safety and environmental effects of our products, processes and waste materials.

8. To work with others to resolve problems created by past handling and disposal of hazardous substances.

9. To participate with government and others in creating responsible laws, regulations and standards to safeguard the community, workplace and environment.

10. To promote the principles and practices of Responsible Care by sharing experiences and offering assistance to others who produce, handle, use, transport or dispose of chemicals.

MANAGING PUBLIC RISK CONTROVERSIES

Obviously, chemical risk issues now are public issues. The following general practices can help foster an atmosphere of trust and responsible progress, especially during controversies.

- Proceed openly, in consultation with affected communities.

- "Frame" the issues sensitively. Because the answers derive from very different sources, distinguish among elements of assessment, appraisal, and response. Encourage the involved and affected parties to articulate their questions and concerns.

- Seek agreement on procedures for resolving any contended technical issues.

- Provide genuine opportunities for citizen participation. Recruit nonadversarial participants to "fill the middle" between extreme views. Involve teachers, physicians, or other citizen-experts in helping frame and interpret the issues.

- Involve civic groups, professional organizations, dispute-resolution organizations, or university groups, to provide expertise and broaden representation.

- As improvements are made in managing risks, inform the public.

ACCOUNTABILITY AND TRUST

This presentation has tried to make the case for robust public discussion of the dimensions of these chemical production risk problems, appraising the risks comparatively, and taking full consideration of the larger social context. These approaches can help reduce social discord and lead to good-faith resolution of the technical and social problems.

Long-term reassurance of the public requires demonstrating two things that may sound like platitudes but that nonetheless are *sine qua non*. First, it must be demonstrated that those "in charge" are competent, honest, well-intentioned, and trustworthy. The proof is in action. And second, it must be demonstrated that improvements are being made -- that accident probabilities and potential consequences are indeed being reduced, the public informed, all due precautions being taken. The showing of progress is reassuring.

A framework for safety in chemical industries in developing countries

Shyamal Ghosh, UNIDO Consultant

D/II-343, Pandara Road, New Delhi-110003(India)

Abstract – There is need to integrate the legislative, administrative and institutional framework for effectively dealing
with safety in chemical production,particularly in a developing country with multiplicity of statutes and agencies dealing
with safety. Lack of an integrated system leads to a frag-
mented approach. Integration can take place under an umbrella
legislation. This will require development of expertise in
Hazard control techniques. The developing countries will need
technical assistance in building up expertise in hazard
control.

INTRODUCTION

The Chemical industry has a crucial role to play in the economic progress of
a country, particularly that of a developing country. But this role is questioned – and rightly so – after any major chemical accident. Consequently,
major chemical accidents have generally led to a total review of all the
systems for ensuring safety in the Chemical Industry.

The components of the systems for ensuring safety can be broadly classified
into the following :

    a) Legislative,
    b) Administrative, and
    c) Institutional.

Each of these components will require to be synergised to enable the development of a well coordinated and effective framework. This paper attempts
to identify the critical inputs of each of these components for promoting an
appropriately integrated regimen in a developing country for containing
hazards and promoting safety.

LEGISLATIVE

A common feature in many legislative systems, dealing with safety, is multiplicity of enactments and their implementing agencies. Sometimes conflicting
signals emanate from such situations because of overlapping of functions of
the administrative arrangements for implementing such enactments. An encouraging trend in recent years has been the adoption of an "umbrella"
approach to coordinate the various aspects of safety in industries, particularly chemical industries. The Occupational Safety & Health Act, 1970(USA)
Industrial Safety & Health Law (Japan) and Health and Safety at Work Act,
1974 (UK) are illustrations of the Umbrella approach. Such an approach is
also now being adopted in developing countries like India.

In fact, this concept of "umbrella"has been further expanded in India through
certain relatively recent legislations. The basic concept of this expanded
approach is that while recognising the need for specialised enactments for
dealing with different aspects of safety (which are not necessarily confined
to the workplace in the Chemical industry), it provides the framework for
coordinating these activities. In India, as in many other countries, there
is a plethora of enactments; the main enactments, prior to the Bhopal
disaster, dealt with safety in Factories/Mines/Ports/Plantations, the governing laws being Factories Act,1948, the Mines Act,1952, the Indian Dock &
Labourers Act, 1934 and the Plantation Labour Act, 1951. There are a number

of other Statutes for regulating safety such as the Petroleum Act, the Explosives Act, the Insecticides Act, the Indian Boilers Act, the Water and Air(Prevention and Control of Pollution) Acts. The Bhopal disaster led to a total review of the situation and the following issues emerged :

a) There was a multiplicity of statutory authorities involved in implementing various aspects of workplace and environmental safety.

b) Each statutory authority confined itself to only a particular segment of safety resulting in a highly segmented approach. For instance, in respect of a hazardous chemical plant, the industrial licence was granted under the Industrial (Development & Regulation) Act - a Central Statute essentially for planning/regulating the development of industries. Gradually, over the years, factors relating to environment were brought into the industrial licensing procedure. But the actual implementation of these factors were to be ensured by the State (Provincial) Environmental authorities. Similarly, factors relating to safety, though not directly related to industrial licensing, had to be taken care of by the State (Provincial) Factory Inspectorate, though under powers flowing from a Central Statute, viz. the Factories Act. Other segments of the factory like Boilers/Electrical Installations/Pressure Vessels, etc. were to be inspected by State or, as the case may be, Central Statutory authorities under the respective Statutes, which were largely Central legislations. Thus a holistic view of any establishment, particularly one manufacturing chemical, was rarely taken in identifying hazards, analysing the risks and reducing them or for that matter for mitigating and containing the effects of a major accident.

c) The intricacies of the Federal-Provincial relationship network in a democratic framework, in the absence of a well-established comprehensive system, further compounded matters. While the bulk of the subjects relating to safety in industry is within the legislative competence of the Federal Government, the execution and administration of these laws is left to the State Government. This underlined the need for establishing, statutorily, certain acceptable standards even though certain model rules/guidelines had been framed.

d) There was need for proper determination of roles of employers/employees/ statutory authorities. A fairly common impression was that granting of any licence by Government (whether it be under Industrial Licensing/Factories Act etc), implied the tacit approval of the State and that the plant proposed to be set up was accepted in its totality, including the safety aspects; this was irrespective of whether all the hazards had been identified by the owner after appropriate risk assessment/analysis and contained. Thus it was recognised that it would be necessary to clearly spell out these rules statutorily.

e) Proper standards/expertise had to be developed for the purpose of ensuring that the safety factor was given due weightage at the pre-project stage itself so that commencing from selection of technology to designing and erection of hazardous installations, the safety factor is built into the system.

f) There was need for coordination of activities under an Umbrella legislation and for strengthening the concerned administrative agencies for establishing a system to control Major Industries Accident Hazards.

A comprehensive enabling umbrella legislation was enacted in India in 1986 which basically provided for (a) legislative cover in dealing with certain aspects of Hazards which had not been very effectively covered earlier; b) a mechanism for coordinating the activities of all the diverse agencies under different statutes concerned with control of Hazards. Side by side, some of the existing legislations were strengthened and it was clearly recognised that the primary responsibility for ensuring safety in any plant was that of the owner, designated as "occupier". A system of regulations was promulgated based on the Seveso Directive (EEC) and CIMAH regulation(UK) for control of major accidents. A clearly defined requirement for safety reports based on identification of Hazards, Risk Analysis, was incorporated in these regulations.

Similarly, steps were taken to improve the administrative and instituional arrangements. The concept of adopting a multi-disciplinary approach by the State authorities was encouraged. Association of outside experts, both by the Chemical Industry and the statutory authorities in identifying hazards and making risk assessments, was also encouraged. In fact, the Government itself set up an Expert Group, consisting of Experts of institutions and Governmental authorities, concerned in Hazard control, to evolve certain guidelines to act as a reference for the Chemical Industry. This Group commissioned surveys and inspection of selected Chemical/Petrochemical units both in the public and private sectors, through specialists, for the purpose of evolving these guidelines, based on observations relating to existing practices in operating units. The report of this Expert Group was comprehensive and included, inter alia, a check list for safety and risk assessments of Chemical and Petrochemical plants. The Chemical Industry, particularly those existing in the small/medium sector could effectively interact with outside experts on the basis of the check list for improving measures for controlling hazards in the units.

The legislative framework, per se, in the context of safety in Chemical Industries has to be comprehensive as well as flexible. As a model, the following conceptual framework could be adopted :

a) A general Umbrella type legislation containing essential features for controlling hazards, coordinating the activities of all related statutory authorities and having the flexibility to promulgate regulations through a process of consultation.

b) Regulations for Control of Dangerous Substances.

c) Regulation for Control of Major Accidents.

d) A standard-setting Mechanism.

The traditional approach of separate laws applicable to, or adopted for certain sectors, from time to time and in piecemeal fashion, will always require constant re-enactment or amendment to deal with the emerging problems of safety and health in the work environment. The umbrella approach has, therefore, to be dynamic as well as flexible to take into consideration modernisation and technological developments. It should be in the nature of an enabling legislation with a view to providing a framework for future detailed or specific statutory duties and responsibilities of all concerned and prescribing standards for occupational safety and health. Some of the essential ingredients of such an approach could be as follows :

a) Evolve parameters about use of substances with which hazards are associated.

b) Recognise general duties and responsibilities of the industry(including both employers and employees), designers, manufacturers, suppliers, importers, etc.,for safety of substances, products or articles used in processes.

c) Evolve standards and codes of practice for ensuring safety.

d) Rationalise existing administrative arrangements to lead to better implementation of laws/practices/standards.

e) Promote and establish suitable forums for consultation and coordination between the law making agencies and the industry.

In respect of the Umbrella type of legislation, there could be a debate as to whether it should be within the purview of the Environmental authorities or the Safety (Factory) authorities. There is a view that since protection of the environment is linked to control of hazards in a Chemical Factory, particularly in respect of hazards leading to toxic emissions, the Environment Protection Act should be the General Legislation to be formulated as a statement of principles. There is also the view that safety is essentially related to the workplace and, therefore, it should be specifically related to Health and Safety in the workplace.

In India, the Environment Protection Act of 1986, which has provided certain enabling provisions for coordinating the activities of all concerned statutory authorities could be termed as a General Legislation under an expanded umbrella concept. For instance, Section 3 of the Act enables the Government to take all measures that may be necessary to protect and improve the environment and includes, inter alia,

a) laying down procedures and safeguards for the prevention of accidents which may cause environmental pollution;

b) laying down procedures and safeguards for the handling of hazardous substances;

c) examining of such manufacturing processes, materials and substances as are likely to cause environmental pollution;

d) inspection of any plant;

e) preparation of manuals, codes or guides.

Further, Section 5 of the Act enables the Government to give directions to "any person, officer or any authority.........". Regulations have been framed in 1989, under this Act, regarding Manufacture, Storage and Import of Hazardous Chemicals. The same rules, with some modifications, have also been suggested for adoption by the State Government as Control of Industrial Major Accident Hazard Rules, 1990 under the Factories Act. This is an apparent duplication but the reasoning for this appears to be that the adoption of such rules under the Factories Act would ensure direct enforcement by the Inspectorates under the Factories Act. However, it will be necessary to exercise caution in cases of such overlap to avoid confusion. In essence, both the sets of regulations broadly follow the CIMAH regulation of UK and the Seveso Directive and require identification of major hazards and steps to contain them, preparation of safety reports and reporting of major accidents.

ADMINISTRATIVE

The Administrative framework is linked to the Legal Structure of a system. Therefore, if the umbrella legislative approach is adopted then logically it should follow that there should be integration in the enforcement network. Individual legislations dealing with various aspects of safety in different workplaces will generally also have its own safety inspection system. For instance, in India there are different functionaries of both Central and State Government acting under the authority of different statutes dealing with various aspects of safety viz - (a) Explosive Inspectors under the Explosive Act, (b) Boiler Inspectors under the Boilers Act, (c) Inspectors under the Insecticides Act, (d) Electricity Inspectors under the Indian Electricity Act and (e) Competent authorities appointed under the Air & Water (Prevention and Control of Pollution) Acts. Multiplicity of inspecting agencies arise due to divided ministerial responsibilities. This may lead to giving conflicting advice/directions. Therefore, an effective way would be to coordinate and rationalise these agencies. The obvious choice would be to have a multi-disciplinary organisation which could deal with various aspects of safety under one roof. For instance, in U.K. such unification of administration has been effected through a separate self-contained organisation with an unified inspectorate set up under the Health and Safety at Work Act, 1974. In India, although the Environment Protection Act, 1986 contemplates constituting an authority (authorities) for the purpose of coordinating the actions of the different statutory authorities and the Manufacture, Storage & Import of Hazardous Chemicals Rules, 1989, notified under the Environment Pollution Act lays down the duties to be performed by various statutory authorities, still no formal unification of all the relevant inspectorates has been effected. The Factory Inspectorates are inducting Chemical specialists as inspectors who, by qualification and training, would be able to meet the challenges of controlling hazards in a Chemical plant. In fact, a multi-disciplinary Major Accident Hazards Control Advisory Division has been established in the Central Labour Institute and Advisory Cells have been established in the Regional Labour Institutes to keep in touch with the State Factory Inspectorates. Similar multi-disciplinary Specialised Cells have been set up by the major State Governments. However, for a developing country where proliferation of statutory agencies has not taken place, the best option would be to create a unified multi-disciplinary organisation under the umbrella legislation.

Even after creating such a unified inspectorate organisation, there is need to built in a mechanism whereby the organisation does not suffer from technological obsolescence. Recruitment of Specialists takes place at a particular point of time in the prevailing technological environment. However, process technology in Chemical factories undergoes changes through innovation/modernisation/upgradation. Therefore, the technological standard of the inspectorates will be required to keep pace with such changes since the techniques of hazard control may require similar changes. Thus, the level of training and re-training will have to be of a very high order.

INSTITUTIONAL

The prime responsibility of ensuring safety in an operating unit being that of the owner, the inhouse expertise to provide necessary inputs must be sufficient and effective. Safety would have to be built into the operational parameters at the design stage itself in respect of a new unit. Certain retroactive measures may become necessary in this regard in respect of an operating unit. There must be a well defined management policy on Safety whereby the critical importance of the role of the safety division is clearly recognised. The safety division should have sufficient manpower and expertise. It is quite possible that sufficient expertise may not be available in the initial stages in a developing country to enable the setting up of an effective inhouse safety organisation.

This aspect has a logical corollary. It may not be sufficient to only maintain a high level of technical efficiency within the operating unit (inhouse) as well as in the statutory inspectorates. It must be recognised that the total talent pool available in the country in this area will have to be utilised for maintaining safety. The experience of many countries indicate that in order to improve safety parameters there must exist technical expert agencies outside Government which are able to give information and advice. Associating professional organisations for specified purposes would help to supplement the inhouse as well as Governmental efforts. However, at least to begin with, it may be necessary to evolve certain criteria for accrediting these agencies/bodies, so that certain minimum standards are maintained. Such accredited agencies/bodies could provide services for inspection and testing. In some countries, these expert professional agencies are statutorily recognised. In Japan, Association of Boilers and Cranes undertake inspections for the employers. Similarly, the Association of Work Environment Measurement undertake periodic testing and measurement of the work environment. In U.K., the Insurance Companies are recognised for certain types of inspections.

In this context, it may be pointed out that in developing countries where the institutional framework for dealing with safety may not have taken sufficient roots, there will be a dearth of such outside experts. Of course, the professional safety consultancy organisation will grow, once the need to associate them in safety audit is recognised. It is in this area that the developing countries will need assistance by way of technological collaboration with the developed countries to ensure that the skills are properly developed.

Apart from existence of expert agencies in the institutional framework, there is also the need for the industry associations to take a lead role in the developing countries. Such associations can provide the forum for pooling of talents for undertaking risk analysis. In fact, in many countries, Chemical Industry Associations have themselves promoted creation of facilities within the Association for undertaking certain specific aspects of risk analysis. This system, apart from ensuring technical excellence also lends critical support to the administrative framework of the Statutory Inspectorates. Such associations also function as a forum for mutual assistance particularly in respect of Chemical Industries in the small and medium sector, where inhouse expertise may not be adequate. Further, the associations provide an excellent channel for exchanging information and build up a good inventory of information. The Industry Associations can also help in compiling Safety Data Sheets which can be gainfully used by all the members. The individual country associations could form a coordinating mechanism for a region or a zone. This could facilitate the transfer of technology with appropriate assistance for analysing risks for identifying hazards and controlling them. This is an area which could form the basis of a project, perhaps, with reference to any particular region. Very often the risks associated with a hazardous process is not properly appreciated by the transferee of technology. Of course, it is necessary that the transferor of technology should assist the transferee in delineating all the hazards in the process and for taking steps to contain them. The concerned Industry Associations could further assist in identifying and controlling the hazards involved in a process, particularly in taking site specific measures.

The other institutional factor involved is the facilities provided at the higher technical educational institutions in the science of risk management. The growth of this discipline requires to be encouraged. In India, only recently, some of these institutions are showing interest in providing special emphasis to techniques of hazard control. Unless these facilities are provided at the University level, it will not encourage the growth of talent in this specialised area and promote the development of organisations specialising in risk assessment techniques. Designing of proper courses to meet the requirements of the Chemical Industry is another area which Industry Associations could help and encourage. Sponsorship of research projects in this area, in the Technical Institution, by the Industry, to keep pace with changing technology, would significantly help in effecting safe transfer of technology.

The role of the UN agencies in promoting and encouraging safety in the Chemical Industry is also an important factor. The ILO has taken up a major project in India for the establishment of a Major Accident Hazard Control Sytem and for training of the factory inspectors. Similar projects have also been taken up in a few other South-East Asian countries. ILO has also published a practical manual on Major Hazard Control. UNEP has initiated studies relating to Hazard Control. It may be appropriate if some of the country projects taken up by individual U.N. agencies could be integrated with other concerned UN agencies so that inputs for such projects become more comprehensive. For instance, the technical input of risk assessment may not be complete unless both ILO and UNIDO work jointly in a Hazard Control Project. Interaction with IUPAC may ensure that the system resulting from such a project takes care of the latest scientific techniques as well as the concerns of the industry.

CONCLUSION

Broadly, therefore, what emerges particularly in the context of a developing country, is that the subject of safety has to be dealt with in an integrated manner in all its aspects. The legislative framework should have a basic enabling legislation which would be comprehensive enough to coordinate the activities of all related aspects of safety and at the same time being flexible enough to take into account changes brought about by modernisation and technological development. Around this basic Umbrella statute, there could be specific enactments to deal with individual functional aspects of safety. Flowing from such a legislative framework there would be an integrated administrative setup to implement the statutory provisions, preferably under one roof. If this was not feasible, the primacy of the umbrella legislation in so far as statutory directions are concerned would have to be clearly recognised. Inhouse safety expertise within the industry would also have to be tailored not merely to meet the statutory requirements but also to fully satisfy the owner that risks have been identified and adequate safety steps taken. Since it will be desirable to involve the entire talent pool within the country for safety, technical and professional organisations as well as Industry Associations will have to be play a key part in setting up an effective and suitable safety regime within the industry. In developing countries, expertise in risk analysis techniques and safety measures may not be adequate. Therefore, mutual assistance through Industry Associations outside the country would be very helpful. International agencies of UN can play a leading role in providing inputs for such purposes at a regional level.

Risk management in the petrochemical industry

Dr. Gh. Ivanus

Process Engineering and Design Institute for Chemical Industry
Bucharest - Romania

Abstract - The paper intends to be a survey of Romanian experience in the field of risk management in the petrochemical
industry as well as to assert the admissible risk. There are
four sequential steps in the process of dealing with risk:
identification, quantification, determination of the risk and
risk reduction measures. Romanian experience is examined comparatively with the advanced world practice in the field, in
a case study related to steam cracking unit. Discussion are
carried about the humanitarian aspect of the measures to be
taken in order to increase the petrochemical plants safety and
their impact on the environment.

INTRODUCTION

Hazard can exist almost anywhere and the avoidance of accidents and their injurious consequences can only be achieved by: firstly, identifying potential
hazards, estimating their significance and, if the circumstances justify it,
in taking steps to improve the situation.

Risk management combines the probability of the occurence of an injurious event
with its consequences and uses the resultant parameters as an aid in the opti-
misation of the use of resources to reduce the probability of event.

There are four sequential steps in the process of dealing with risks: identifying the dangers to people or the environment; quantifying the extent of
these dangers; determination of the acceptability of the risk; risk reduction
measures.

This strategy has been applied to the fields of the petrochemical industry as
we explain below.

IDENTIFICATION

In this step it is established whether a petrochemical plant can cause a hazard to its surroundings and the nature of the hazard is defined.

In the study of the base case carried out, we analysed the events occured in
the petrochemical sector in six (6) big petrochemical complexes sites, three
petroleum refineries and a carbon black works.

By events the study conducted means: short time shut downs due to mechanical
failures, errors of operation and primary and secondary energy supply, small
fires during one operation year. No major events have been taken into account.
The distribution of the events studied on the site of 10 industrial units is
shown in Table 1 (6,7,8).

A first conclusion to be drawn from the analysis of the data in Table 1 consists in the high frequency of events occured in the petrochemical sector
compared to the number of events occured in the refinery sector, during the
same period of reference, justifiable if we think to the complexity of petrochemical processes.

TABLE 1 - Distribution of Events on the Site

Unit	Frequency (number of annual events)	Severity (loss by event express in US dollars)	Density (number of events by 100 ha)
Petrochemical (1)	30	26,000	6.8
Petrochemical (2)	25	40,000	5.10
Petrochemical (3)	19	35,000	4.8
Petrochemical (4)	10	42,000	3.0
Petrochemical (5)	6	28,000	0.6
Petrochemical (6)	4	23,000	0.5
TOTAL ITEM NOS. 1-6	94	-	-
Refinery (7)	8	25,000	3
Refinery (8)	5	20,000	2
Refinery (9)	2	18,000	0.5
Carbon black works (10)	1	24,000	0.2
TOTAL ITEM NOS. 7-10	16	-	-
TOTAL ITEM NOS. 1-10	110	-	-

The main causes leading to the occurrence of such events can be classified as follows:

- Unsuitable mechanical maintenance   40 events
- Noncompliance with the operation instructions   24 events
- Abnormal supply of utilities (steam, water, air,
  nitrogen)   20 events
- Power failure (electric power)   16 events
- Uncontrolled welding operations   5 events
- Spontaneous combustion or explosions   5 events

                                            TOTAL   110 events

It is worth underlying the significant share of events due to mechanical revisions carried out unproperly (lack of spare parts, revision period shortened, use of unsuitable materials, steels) operation errors and abnormal supply of utilities. A frequent cause of emergencies in the petrochemical sector is represented by disturbances occurred during pressure vessels operation. The study under discussion analysed 140 emergency cases at vessels with flammable products under pressure, emergencies identified as follows: visual examination 60%, pressure tests 30%, nondistructive inspection 6%, consequences of explosions 4%.

In about 90% of the emergency cases there have been found cracks near welds in the vicinity of nozzles attached to the cylindrical part of vessels and over 50% of the total emergencies cases at pressure vessels are due to failures of other types of petrochemical equipment is shown as follows (see Table 2) (4)

TABLE 2 - Equipment Failures

Type of Equipment	Percentage Share (%)
1. Rotary machines	35
2. Heating and reaction furnace	12
3. Heat exchangers	14
4. Pipes	9
5. Columns, vessels	8
6. Instruments	7
7. Electrical equipment	6
8. Boilers	5
9. Storage tanks	4
TOTAL	100

QUANTIFICATION

Once the hazards have been identified, the next step in risk assessment is to quantify the risk associated with each hazard. This is accomplished by determining two parameters:

- the annual frequencies of an event, and
- the corresponding severity of the accident

The product of these two parameters for any hazard is a measure of the annual risk (5)

$$\text{Risk} = \text{Frequency} \times \text{Severity}$$

| (expected loss per year) | (expected no. of events per year) | (average loss per event) |

Table 1 shows the elements necessary for risk assessment, namely the frequency and severity for the base case study conducted by us in the petrochemical sector and several petroleum refineries during one calendar year.

Risk quantification by means of expression (1) needs a classification of previous emergency accidents in order to establish the corresponding annual losses and therefore to estimate the average annual loss (that is severity) for each event separately and to preestablish the frequency of emergency, that is of the probable number of accidents. Such classification and estimation can be done only on the basis of existing data stored during the operation years of plants making the object of this study or of similar plants.

ACCEPTABILITY OF THE RISK

There are some general criteria (C_1, C_2, C_3) for acceptable risk determination, such as (5) :

C_1 - decomposition enthalpy criterion $\triangle H.d$

when $\triangle Hd >$ - o.7 Kcal/g there are high risks

when o.7 $<$ $\triangle Hd$ $<$ - o.3 Kcal/g there are acceptable risks

when $\triangle Hd <$ - o.3 Kcal/g there are negligeable risks

C_2 - combustion enthalpy criterion $\triangle Hc$

when $\triangle Hd \simeq \triangle Hc$ there are high risks

C_3 - functional groups and unstable structures

The technical literature gives the groups considered as unstable due to their high reactivity, such as:

$- C \equiv C -$	acetylenes	$>C - NO$	nitrocompounds
$>C - O - O - C<$	hydroperoxides	$>C - N - N - C$	azocompounds
$- C - O - O - H$ (with O below)	peracids	$>CN_2$	diazocompounds
$>C - C<$ (with O below, epoxide ring)	epoxides	$- N = N - N = N -$	tetrazols
		$>N - NO_2$	nitroamines

C_1, C_2 criteria and C_3 respectively are of a nature to give a general idea over the hazardeous degree of some process equipment items wherein chemical reactions take place without nevertheless contributing to their quantification. Due to this reason the following method are somewhat more specific to the studied field in view of acceptable risk assessment.

The experience proves that any industrial activity so much more the chemical and petrochemical industry assumes the existence of a certain amount of risk which can be eliminated only during the total shutdown of the plants in question. For this reason one can discuss only about the reduction of the industrial risk to a certain degree either by means of additional investments provided in the initial design or subsequently by commissioning of petrochemical plants.

Experience also proves that the reduction of industrial risk is not proportional to the increase of investments done, so that from a certain investment level, the risk level is virtually constant. For this reason specialists raised the question of ascertaining an "acceptable industrial risk" beyond which the additional capital investment concerning operation safety is no longer justifiable.

Fig 1 Acceptability of the risk

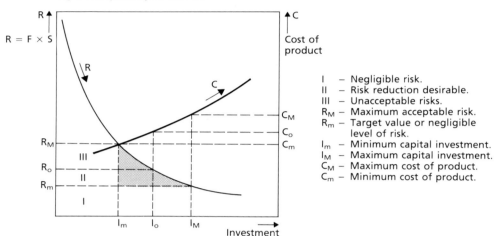

Fig.1 shows a more complete method representing the variation of basic parameters involved in the assessment of acceptable risk id est: invested capital for risk reduction, production costs and the size of resulted industrial risk

Determination of the pairs of RM Im, Rm IM, IM CM and Im Cm is a difficult enough operation differing significantly from one kind of petrochemical plant to the other and often from one site to the other of the same kind of plant depending on local existing conditions.

The subsequent selection of optimum risk assumed to which the pair of Io and Co values corresponds depends to a great extent on the experience of those in charge with "risk management" and on the option of the administrative board of the Company under study.

The use of any measures aiming at improving the technical safety of petrochemical plants cannot surpass some reasonable "possibilities" or limits. For instance one cannot agree with any expense on the industrial risk reduction by additional investments as they are implied in the production costs and the profit level (see Fig.2)   (4)

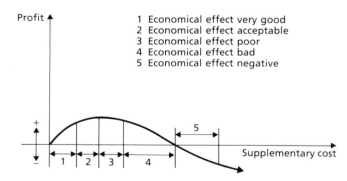

Often the level of concern that a company has for some types of accidents is greater than that for other types of accidents. For example, a company may be willing to spend resources to reduce the high-consequence accidents (low frequencies, high severity) while they may not be willing to spend more to prevent nuisance type of accidents (high frequency, low severity) even if the absolute risk associated with the two types of accidents is the same.

The maximum acceptable individual risk can be established in the range of 10^{-6} per year. In other workds the risk of a fatal accident to which an individual is exposed because of his continuous presence (365 days per year) in the neighbourhood of an activity with dangerous goods shall be less than once in a million years.

For the maximum acceptable group risk level chance of 10^{-5} per year of an incident with maximum 10 deaths has been chosen.

RISK REDUCING MEASURES

Any study dealing with the "risk analysis concept" is useful only to the extent it recommends some efficient measures for reducing the industrial risk in view of protecting the operation personnel, the people living in the neighbourhood of petrochemical plants and diminishing the volume of technical damage leading to production costs.

In order to prevent some severe accidents due to gas or light hydrocarbons leaks it is necessary to quickly detect them by means of automatic explosion-meters equipped with detectors imediately detecting the presence of gases which might make explosive mixtures. Explosionmeters can detect the presence of explosive gases when in concentrations below the lower explosive limits i.e. below 30% of the low explosion limit (LIE) specific to each substance separately. This makes it possible to detect the presence of explosive gases in certain plants areas long before they become dangerous.

The base case study conducted by us and localized in a steam cracking unit determined the concentrations of flammable gases (ethane-pentane) corresponding to prealarm and alarm limits as shown in Table 3.

TABLE 3 - Concentration of C_2-C_5 Hydrocarbons   (6)

Process area of pyrolysis plant	Alarm set off concentrations (% of LIE)	
	Prealarm	Alarm
Pyrolysis furnace	30	60
Cold - box	30	60
Hot area	50	80
Peripheral area outside the plant	30	70

In case the prealarm limit is set off in a certain area of the plant, the operator on duty shall check by means of portable explosionmeter the presence of gases opposite to wind direction as shown by gyron and informing the shift foreman about the conditions found.

When the alarm is set off steps shall be taken for action or shutdown as there is a risk for the flammable gases to reach the explosive limits.

Based on the economic risk analysis we developed some specific recommendations for the individual process unit existing in petrochemical units or refineries.

C o m p r e s s o r s : Accidents involving process sections containing compressors are the dominant economic risk contributors more than 35%.
We recommend that the petrochemical units maintain a regular surveillance inspection and testing program for these large compressors, vibration detection and incipient disruptive compressor failures which can lead to loss of containment accidents.

E n c l o s e d   a r e a : a release of light hydrocarbon within an enclosed area creates the potential for explosions. Therefore we recommend that the petrochemical units must provide the following: continuous, positive ventilation in such areas especially in compressor buildings, when local climatic conditions allow, compressor units can be installed in open area, control room alarms and compressors shutdowns on high explosive levels in such areas

H e a t e r s : process heaters are the second most important component type, contributing almost one-fifth of the absolute risk, therefore we recommend that petrochemical units need to institute a reliability improvement program for process heaters.

P u m p s : the most common type of pump loss of containment event is a seal leak or rupture. Monitoring the performance of pump seals, especially for heat pumps, will allow for detection of incipient failure and the reduction of the expected frequency of leaks. Periodic use of vibration detection equipment will allow the petrochemical units to detect incipient internal pump failures such as bearing failures or impeller rubbing that could cause pump ruptures.

CONCLUSIONS

1.   As compared to the methods published by technical literature, the present paper suggests, as a way of acceptable risk determination, their correlation to the investment costs and production costs. Correlation of additional benefit and safety technique costs is also an idea belonging to this paper.

2.   Among industrial risk reduction measures in the petrochemical sector according to a base case study conducted for steam cracking unit a special part is played by installation of explosive gas detectors controlled in such a way as to energize prealarms and alarms when the concentrations reached are below the lower explosive limits of C_2-C_5 hydrocarbons, ethane-pentane.

3.   For the time being the engineering activity does not have a "universal" method while the administrative boards do not even have an "official" one for the quantification of acceptable technical risk.

After having experienced some unhappy events, the competent authorities appointed expert teams to investigate the causes although the technical risk is still not zero but eventually diminished only down to a so called "acceptable conventional" level.

BIBLIOGRAPHY

1.   S.P. Maltezou, K.Biswas and H.Sutter; Hazardous Waste Management. Edited Tycooly 1989

2.   C.J van Kuijen, Risk management in the Netherlands; A Quantitative approach: International Expert Workshop, UNIDO 22-28 Jun.1987

3.   Niels K. Vestergaard: Risk assessment, Danish practice and experience International Expert Workshop, UNIDO 22-26 Jun.1987

4.   A.Pavel: Siguranta in functionare a utilajelor petrochimice. Editura Tehnica Bucuresti 1988

5.   C.Carloganu, Combustii rapide in gaze si in pulberi. Editura tehnica Bucuresti 1986

6.   Gh.Ivanus, Hazard and risk assessment manual - Printed by UNIDO 1986

7.   Gh.Ivanus, Criteria for risk assessment and overview of International regulatory aspects for safety provisions

8.   Gh.Ivanus, Reduction of the Industrial Risk for Chemical Technologies World Conference on Industrial Risk Management and Clean Technologies Nov. 13-17, 1988, Vienna

9.   M.Renert: Fiabilitatea utilajelor si instalatiilor industriei chimice Editura tehnica, Bucuresti, 1980

10.   R.Nichols: Pressure vessel engineering technology. Elsevier Publishing Co., Amsterdam, 1971

Oil field disaster management
An approach

Dr. S.M. Sharma

Chief Chemist, Safety & Environment Management, Oil & Natural Gas Commission, DEHRADUN (UP) - INDIA.

INTRODUCTION

Some of the major industrial disasters occurred in the recent past have greatly aroused the public opinion and attention of National and International regulatory agencies. These events include those which are naturally occurring, such as the:

-        Earthquake that struck Mexico City in 1985, and Armenia and Khazikistan (USSR) in 1988.

-        Mud slides in Ecuador in 1987

-        Release of toxic fumes from a lake in Cameroon.

They also include industry-related disasters such as the dioxin-containing release in Seveso, the propane explosion in Mexico City, the release of Methyl iso-cynate (MIC) at Bhopal, the fire and discharge of contaminated waters into the Rhine from a warehouse in Basle or the release of radio-activity from a nuclear reactor in Chernobyl.

Science has not progressed to the stage where all the causes of events leading to Disasters both natural and man-made, are understood, predicted or effectively prevented. In the meantime, there is need to prepare ourselves to respond to these emergencies, as and when they occur. Safety experts in Industry, on the other hand, have the philosophy that all industrial accidents are preventable. Yet they are realistic enough also to prepare Disaster Management Plans should such accidents occur.

MAJOR HAZARD

As on today, none of the legislation in Europe or India defines a major hazard. Definition of hazardous installations, processes, storage levels beyond which an installation can be deemed hazardous are available in the Environment Act, 1986, Factories Act, 1987, EEC Direction 1984 and CIMAH regulation, 1984. Major hazard is one in which the health and well being of the community is affected, social life is disrupted temporarily or permanently. This disruption could be a result of:

A   Occurrence of an explosion, release of gas, toxic or nuclear materials, liquid spills, vapour cloud etc. These create a situation of emergency and considerable effort has been spent in the recent past to deal with such situations.

B   An insidious process which goes on gradually over a period of time.

After the Bhopal tragedy and others, our attention in India has unfortunately been focussed on the (A) above and Nelson's eye has been turned towards (B).

RESPONSIBILITY IDENTIFICATIONS

Basics

Control of major Hazards involves a triangle of :

-        Those who, due to lack of control or otherwise, create hazards (Creator).

-        Those whose job is to protect the community (Protector).

-        Those whose function is to clean up the affect of the hazards (Normaliser).

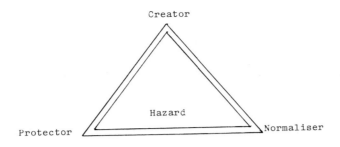

Hence, control of hazards would require an interaction and close co-operation of all the three agencies, if this system is to work. If one or more agencies of the above triangle does not perform or perform inefficiently, misery of the community at large is inevitable.

OIL FIELD DISASTER PLANNING IN ONGC

Oil & Natural Gas Commission (ONGC), a Government of India Undertaking, is responsible for exploration and exploitation of hydrocarbon resources in India. ONGC's main objective is self suffiency in oil. Oil is providing fuel and feed stock to power plants, mills, refineries, fertilisers, petrochemicals and other petroleum based industries beside domestic gas and LPG.

ONGC's drilling and production activities involve some degree of risk, accidents and other dangerous/ disastrous situations like :

- Blow Out
- Fire
- Oil Spill
- Emission of H_2S and combustible gases
- Helicopter crash on deck etc.

During such disasters, for the safety of personnel, offshore rigs are built under strictest quality control supervision of shipyard owners and classified society like American Bureau of Shipping etc. Government enforces rules and regulations of Safety of life at Sea (SOLAS) and all the necessary life saving equipment like life boat etc. are provided. Trained personnel are deputed to operate these sophisticated rigs and knowledge of dangers and safety training is imparted to persons working so that emergency situations could be controlled without panic. But still in spite of all these, things go wrong and emergencies do arise involving damage to life and property.

CONTINGENCY PLAN

A contingency plan is a pre-planned procedure for handling emergencies with a view to minimise loss of life, property and damage to environment and bring the situation under control, as soon as possible. Disaster planning needs fixing up of prorities in following order:

- Safety of life
- Protection of installations and environment
- Restoration of Production.

Once the persons are rescued, action should be taken to limiting pollution, spills, controlling emergency situation and finally restoration of normal operation as they were prior to disaster.

A   BASIC STRUCTURES OF CONTINGENCY PLAN

i)   Action Plan indicating duties : starting from site of crisis, raising alarm, communication to each member of contingency plan.

ii)   Comprehensive guidelines to all concerned.

iii) Information about equipments, services and other support facility required along with contact addresses and full specifications of equipment.

iv)   Chain of command and responsibility.

B.   CRISIS MANAGEMENT

i)   Comprehensive guidelines are given to all concerned.

ii)   Overall supervision, guidance, expertise and special stores and equipments, additional trained manpower and material resources and timely replacement of these in all areas of shortfall.

iii)   Secure maximum efficiency and economy in implementation of plan.

iv)   It is neither feasible nor proposed to build up administrative capability, stores and infra structural facilities for responding to an infrequent irregular and unforeseen crisis situation, rather it would be more prudent to work out arrangement whereby available administrative capabilities can be mobilised within shortest possible time with maximum effect under unified command and control structure as it is done in war time.

v)   Besides the response parts for different crisis situation, plan contains many common elements such as communication, transport, regulatory rescue and relief measures, medical and hospitalisation as such it is admissible to plan out administrative arrangements and all common elements for various envisaged situations.

vi)   All special requirements for specific crisis situation can be separately formulated and thereafter amalgamated as specific appendixes to the basic plan.

vii)   Formulation and implementation of such integrated plan would, however, require close interaction and cooperation of various concerned departments and sections of administration.

viii)   Listing of identified emergency personnel and clear chapter of duties of them. The plans are drawn and updated from time to time.

ix)   Briefing the media/press for circulation of correct information so as to allay apprehensions and prevent spread of rumours which may cause panic elsewhere.

x)   Establishment of well equipped control room at site/near to site, base and quick assemblage of all the essential team members at the control room to assume charge of the situation. The essential information for taking steps should be readily available in plan document in control room.

C.   CONTROL ROOM

i)   The main control room should work as nerve-centre of all emergency operations and hence should be adequately constructed/located and equipped with arrangements of round the clock normal message centre with skeleton staff which can be converted into full scale operation room once a crisis has arisen.

   The duty officer should have in possession names, addresses and telephone numbers of all the members of crisis management; communicate till the arrival of regular functionaries. Control rooms in addition to adequate staff should have sufficient telephones, wirelesses, radio communication/hotline installed in advance and maintained in efficient condition.

   Site Control Room should be set up nearest to the accident site to assess the situation of site and timely apprise main control room.

ii)   Chief Emergency Coordinator (CEC) : Main task of CEC is to assess emergency situation, take suitable action on priority to protect life, damage to property and environment and finally bring accident site into normal working place as it was prior to accident. He must visit site of accident, arrange all stores and services required, inform situation to top management, Government authorities, media, local authorities like Police etc., if required. He must keep contact with site emergency coordinator.

iii)   Site Emergency Coordinator (SEC) : Top level person present on the site must take charge of situation and command all operations. Declare emergency, if required. Inform nearby installations and energise main control room. He should organise evacuation for safety of persons and take other steps to minimise damages to property/installations with available means at his disposal.

iv)   Key Personnel : Chief Coordinator, Incharge of Main Control Room, should be assisted by the senior managers of technical, scientific and administrative wings. These persons have key role to play in advising CEC and SEC in taking appropriate decisions and implement the same. No sooner emergency is declared, they must assemble at Control Room.

v)   Drill Practice : The regular drill/rehearsal will help emergency personnel in remembering details of their assigned task and enable them to reach instantaneously in actual crisis situation.

DISASTER MANAGEMENT ON OFFSHORE RIG - A BLOW OUT CASE HISTORY

A.   A blow out is the result of a sudden uncontrolled release of gas or gas-oil at extremely high pressure. This unexpected high pressure release occurs at great velocity and the large amounts of material evolved are at a high temperature. Hence, the risks of explosion, fire and wide-spread pollution are all present. Blow outs represent the potential loss of life and equipment, as well as loss of the gas or gas-oil itself, together with the associated revenue and energy content.

Blow outs can occur on drilling into an unexpectedly high geological pressure zones, from human error, equipment failure, or external causes such as collision or fire. The possibility of a blow out exists at all drilling stages, during production, and during maintenance and inspection of the drilling and production equipments. A proposed model of action in case of blow out is given at Annexure I and II.

B.   CASE HISTORY

-     ONGC owned mobile offshore jack up drilling rig commissioned on 1.1.82, deployed on platform "SG" in Bombay High 150 Km. from shore, equipped with life saving, fire fighting and blow out control equipments.

-     July 30, 1982 at 9.15 PM Blow out occured. All the 74 persons on board were evacuated within half an hour to safer place 5 Km. away.

-     At 9.20 PM, Member (Offshore) received the information on phone. Within minutes, he was in Control Room at Maker Towers directing operations. Fire fighting boats arrived within 4 hours and started spraying water jet over threatened rig.

-     World renowned blow out expert 'Red Adair' was connected in Houston on phone.

-     Member (Offshore)/GM flew by night helicopter to nearest location to assess and battle the hazardous situation.

-     July 31st, 1982 at 6 AM Chairman was informed at Delhi. By 10.30 AM he was at the site.

-     Another MSV Pacific Constructor equipped with powerful fire fighting pumps positioned 500 mts. awaya from SJ platform swing into action of spraying water. Site control room was shifted to MSV Pacific Constructor.

-     ONGC Management identified the priorities :

        -     Contain blow out, fight fire, mobilise resources equipment, and expertise from all over the world.

        -     Keep moral of staff high.

        -     Ensure participation of media.

-     August 1, 1982 : Helicopter landed on top of front leg of rig, a man got out and switched off running generator and returned by the same Chopper.

-     August 2, 1982 : At 6.17 AM inspite of all precautions taken, Well SJ-5 caught fire. Fire flames 40-50 mts. high started destroying metallic structures and most of the equipment fell into sea.

-     Gulf Fleet 46 was moving around the rig and spraying water.

-     7th Aug., 1982 : another fire fighting boat Gulf Fleet 47 was mobilised from Madras and MSV FLEX SERVICE with high capacity fire fighting equipment came from Dubai on 8th August, 1982.

-     3rd August, 1982 : At 4.44 PM the fire died down on its own but the blow out continued.

-     To drill releif well semi submersible rig arrived on site on 14th August, '82.

Annexure: II

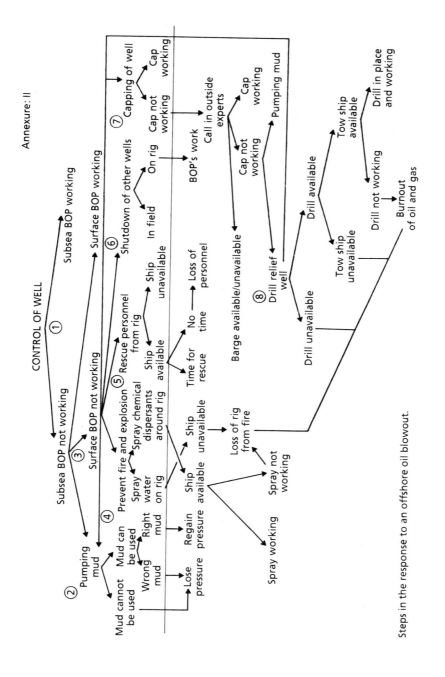

Steps in the response to an offshore oil blowout.

Annexure: II

Proposed model of action in case of flowout

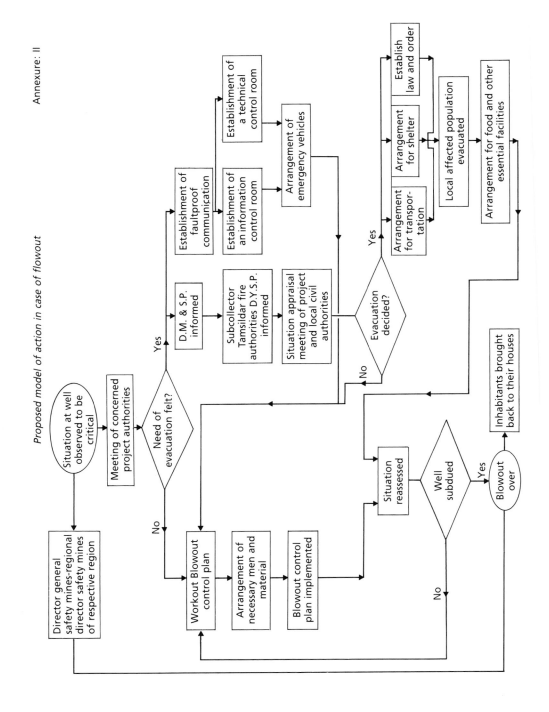

- Naval ships/air crafts, Air Force helicopters and Coast Guard vessels assisted in controlling pollution. No pollution was caused by this blow out.

- Chairman addressed the nation on T.V. for 10 minutes on Blow out and steps are being taken to control the blazing well.

  Derrick Barge HERCULES arrived on 18th Aug.'82 and was stationed adjoining to SJ platform. Crawler Crane was lifted by giant crane of Hercules and placed on the rig for assisting capping operations.

- Helicopter with ONGC officials and Red Adair team landed at Rig to plan capping operations. On 27th Aug. '82 Hercules started clearing debris of SJ Platform and on 29th Aug. '82 top deck.

- 30th Aug. '82 : Well head and BOP was installed for capping and killing operations.

- 12th Sep. '82 : At 1140 AM well was closed and killed, blow out situation was over.

- Damaged rig was jacked down and towed to Bombay harbour, sea bed, area around SJ platform was cleared of debris. Another rig was brought and oil producer well was successfully drilled.

C. OUTLOOK FOR FUTURE

As a result of lesson learnt from this blow out, following modifications have been done in new rig :

- Helicopter landing pad on top of leg has been provided.

- Remote control of stopping emergency generator from life boat embarkation station has been provided.

- Fire monitors for spraying water of rig floor has been provided.

CONCLUSION

- With a view to minimise loss of life, damage to property and environment and to bring the situation under control as quickly as possible, contingency plan must be updated after learning lessons from actual incidents or drills.

- A good communication network, trained personnel, specifically required equipment and laid down emergency procedures are main key points for effectiveness of any contingency plan.

. ACKNOWLEDGEMENT

Author is highly indebted to Chairman, ONGC for his kind permission to present this paper in the workshop at Basal, Switzerland. He is grateful to Shri S.K.Manglik, Member (Tech.) for his advise, encouragement and approval of the paper. Author had also useful discussions with Shri G.B.Sharma, GM (SEM). His guidance and support is thankfully acknowledged.

REFERENCES

1. Recommended Code of Practices, ONGC Publication, 1985.

2. Blow Out Contingency Plan, ONGC Publication.

3. Offshore Emergencies Handbook, UK, Deptt. of Energy, Petroleum Engineering Division, London, Nov. '84.

4. Offshore Installations (Emergency Procedures) Regulation, UK, 1978.

5. Guidelines for Industrial Contingency Planning. International Maritime Organisation, MEPC/CIRC 122, London, Feb. 1984.

6. Risk Management Concept on Fire Loss Prevention, S.P. Bag SAIL Durgapur, 1986.

7. Disaster Management, Dr. A.Ahmed, ASCI, Hyderabad, 1981.

8. Managing Technological Accidents : Two blow outs in the North Sea, D.W.Fiacher Pergemon Press, 1982.

Day 2: Case Studies

Day 2: Case Studies

Workshop participants joined one of twelve case studies led by a member of the Basle Chemical Industry. This gave a practical opportunity to personally consider the safety aspects of a process and the safety assessment method used to minimise risk.

The twelve processes investigated and the team leader of each investigation were:

Nr	Topic	Process Description	Major Safety Problems	Risk Analysis Method	Leader
1	Catalytic hydrogenation	Transformation of the triple bond of C30-dehydroaldehyde into an olefinic double bond by means of catalytic low-pressure hydrogenation in a 10,000 litres multi-purpose hydrogenator.	– Fire and explosive hazards – Inertization, gas exchange – Handling of pyrophoric catalysts – Process control by computer	Application of a screening method on process, equipment and environment, by using check-lists.	Dr K Meyer Dr M Septinus F Hoffmann-la Roche AG
2	Nitration	Production plant for several hundred tons of nitrated benzene-intermediates for various dyestuffs using concentrated sulfuric acid as solvent (start-up 1977). Nitrations of organic materials are potentially dangerous processes due to the highly exotherm ic reaction during the addition of mixed acid (50% nitric, 50% sulfuric acid). In order to achieve the goal of building the safest possible plant it was decided to design a continuous production plant in the form of a cascade.	– Exothermic reaction – Thermal stability of reaction mass	'Zurich'-Hazard Analysis	P Meyer Sandoz Products (Switzerland) AG
3	'Marine-Blue'	On the example of a dye stuff manufacture including a diazotization reaction, the different steps of a systematic risk analysis of a chemical process are demonstrated. 2-Chloro-4,6-dinitroaniline dissolved in sulfuric	Thermal stability of diazo mass; exothermic reaction.	The process will be analyzed for risk with a systematic checklist method (suitable for the investigation of batch processes).	F Altorfer CIBA-GEIGY AG

Nr	Topic	Process Description	Major Safety Problems	Risk Analysis Method	Leader
		acid is diazotized with nitrosyl sulfuric acid in a 2,500 l glasslined vessel. Due to the very exothermic reaction and thermal instability of the diazo mass, special attention has to be given to the temperature control during the addition of the nitrosyl sulfuric acid.			
4	Catalytic Hydrogenation of 2,4-Dinitro-chlorobenzene	2,4-Dinitrochloro-benzene dissolved in methanol is hydro-genated with Raney-Nickel as catalyst in a 'Buss'-Loop Reactor. The addition of hydro-gen is defined as 'pseudo steady-state condition' (as in a continuous reactor) and will be analyzed for risks with the 'Hazard and Oper-ability Study' Method.	– Thermal stability of reaction mass – Handling of hydrogen and catalyst	'Hazard and Operability Study' (suit-able for the investigation of continuous processes).	Dr R Grunder CIBA-GEIGY AG
5	Sulfonation of a nitro-aromatic compound	A nitro-aromatic compound is dissolved in concentrated sulfuric acid. The solution is heated to the reaction temper-ature (90°C) and an excess of oleum (154%) is added to the solution over a period of 10 hours. After the addition, the reaction mixture is agitated at reaction temperature for 9 hours to complete the reaction.	– Thermal stability of reaction mixture – Accumulation of reactants	Calorimetry and numeric simulation. (Aim of the case study is to show how safety aspects can be integ-rated into the development of a process and lead to a safe and economic operation).	Dr F Stoessel CIBA-GEIGY AG
6	Explosion protection in a 'Grinding Unit'	Each unit part or unit operation is analyzed in detail for special hazards that could occur as a result of equipment and operating errors. An overview over the possible technical explosion safeguards and their application is given. Finally the chosen protective measure 'Explosion Pressure Venting' is discussed in detail and their adoption to the partic-ular local situation as required by the technical safety data of the products being ground	Explosion protection		R Siwek CIBA-GEIGY AG

Nr	Topic	Process Description	Major Safety Problems	Risk Analysis Method	Leader
		is described.			
7	Explosion protection of a spray drier in combination with a fluid bed drier	A spray drier in combination with a fluid bed drier and a dust separator has to be protected against the effects of a possible dust explosion. The different methods of protection which in principle can be applied are discussed. Finally the measure 'explosion suppression' is selected and discussed in detail.	Explosion protection		Dr M Glor CIBA-GEIGY AG
8	Fluid bed dryer in pharmaceutical production	Drying of granulates in pharmaceutical production by means of a fluid-bed dryer.	– Formation of explosible mixtures of dust and vapours ('hybrid' mixtures) with low minimum ignition energy – Explosion protection	Application of a screening method on process, equipment and environment, by using check-lists.	Dr K Aeberhard F Hoffmann-la Roche AG
9	Waste water drains	The waste water from different buildings is collected in a main drain and brought to a pump station, from where it is conveyed to the water treatment plant. If by accident solvents get into the water this might lead to an explosion.	Explosion, destruction of pipes	'Zurich Hazard Analysis'	W Zenhäusern Sandoz Pharma AG
10	Risk analysis of a high rack warehouse	Object of the study is a high rack warehouse with a high degree of mechanization and automation for the storage of chemical raw materials, intermediate substances and finished products.	– Fire detection, fire protection and fire fighting, including associated problems (eg. fire water retention basin) – Classification and segregation of stored goods – Handling, labelling, identification of goods	'Zurich Hazard Analysis'	Dr H Rudolf Sandoz Produkte Schweiz AG

Nr	Topic	Process Description	Major Safety Problems	Risk Analysis Method	Leader
11	Chemical exhaust-air cleaning unit	Construction and operation of a chemical exhaust-air cleaning unit (existing pilot-plant) with subsequent incineration (in planning phase). Collection of exhausts out of different multipurpose reactors.	Prevention of cross-contamination, suppression of fire or explosion backlashes; control of inherent hazards of incineration.	Application of a screening method on process, equipment and environment, by using check-lists.	Dr J van Koeveringe F Hoffmann la Roche AG
12	Occupational hygiene	In a chemical production plant, the problems of occupational hygiene will be addressed. - Hazards will be evaluated based on process and plant documentation by evaluating . the properties of substances, where data can be found or how they can be generated . the monitoring methods . the exposure situation - Work place analysis and monitoring will be carried out - The results will be evaluated and - measures discussed which may need to be taken.	Occupational hygiene		Dr R Roth Sandoz Technologie AG

Day 3: Safety Education Today

Introductory remarks

[1]Introduction to Panels 4-9

Lothar Reh, ETH Zürich/Switzerland

Chairman Day 3

The program of day 3 is completely devoted to the evaluation of today's safety education at universities, engineering schools, industry and authories in different countries of the world. A special focus is directed to the situation in developing countries.

Under the guidance of experts from universities and industry the present status of teaching in different fields of safety precautions will be discussed in the panels 4 to 9.

It is the intention of the organizers, that recommendations will be elaborated for instance

- how the different safety features can be optimally integraded into teaching programs of different schools and external courses,

- how much safety training is feasible without overloading undergraduate programs,

- how far can safety training be improved by practizing it in university basic training and research laboratories, and

- how safety training of plant operators is optimally performed.

Beside discussing these and possible other questions, the mode of teaching safety precautions will be demonstrated by typical case studies, which can be adapted to different levels of education.

All case studies presented in the panels are distributed with this volume to all participants of the IUPAC-Workshop "Safety in Chemical Production". They are examples, how safety precautions can be taught successfully. Everybody teaching in the field can use them free of copyright for inclusion into lecturers and courses.

To facilitate the access to important safety training material, Mrs. Sue Cox from Centre for Extension Studies, Loughborough University of Technology has elaborated on behalf of the organizing committee of the present IUPAC-Workshop a comprehensive bibliography.

Special thanks are given to Mrs. S. Cox for undertaking this valuable task and to all panel chairmen for their careful preparation of the case studies.

Panel 1: Safety Education at Universities

MODERATORS
D. A. Crowl and T. A. Kletz

Safety education at the university

Daniel A. Crowl

Dept. of Chemical Engineering
Wayne State University, Detroit, MI 48202

Abstract – In the last several years chemical process safety has become very important to industry in the United States. This has resulted in a mandate for additional chemical process safety instruction at American universities.

This paper will describe an extensive program in the United States to include additional safety instruction at the universities. This program features a number of aspects, including 1) changes in accreditation requirements, 2) development of necessary instructional resources, and 3) training of faculty in this new subject area.

Currently, the implementation of this program is restricted in that present chemical engineering curricula are overloaded. A number of methods are suggested to solve this problem.

The program currently underway in the United States could be implemented in other countries.

HISTORICAL BACKGROUND

During the summer of 1986 I enjoyed an opportunity to work in the Chemical Engineering Department of BASF Corporation in Wyandotte, Michigan. One day the director of the department came to my office and asked me if I was interested in working on a project related to process safety. In total ignorance I replied "You mean hard hats and safety shoes?" Little did I realize that this response reflected the common academic ignorance of process safety technology at that time. Fortunately, my host was able to show that a large area of fundamental technology existed which was mostly unknown in academic circles. This area of process safety technology had been developed by industry during the past twenty years mostly in response to increasing needs in this area. At the completion of my summer visit, I had become aware of runaway reactions, two-phase flow through reactor reliefs, fires and explosions, and a whole host of fundamental technology which I had previously known little about.

Prior to the Bhopal tragedy in 1984, most chemical companies were making steady progress at increasing the technology and management of industrial safety programs. This was done almost entirely without any academic involvement. After Bhopal, public outrage insisted on a new mandate for safety at a level of activity much higher than before. As a result of this new mandate, industry began to insist on more safety instruction at American universities.

At that time American universities were in a difficult position to respond to this new mandate. First, faculty had little expertise in this area or had industrial experience during a time when safety was given less emphasis. Second, the engineering curriculum was overloaded. Which material should be deleted to include more safety instruction? Finally, no suitable instructional materials were available to teach the necessary courses.

A survey (ref. 1) of industrial and academic individuals determined that the problem is compounded by a completely different perception of safety and health between the academic and industrial communities. The results are illustrated in Figure 1. They show that the academic community perceives safety to be mostly chemical hazards and ethics while industry perceives safety in a more technical fashion, including risk assessment, runaway reactions and plant practices and procedures. Only in the area of fires and explosions was there reasonable agreement on relevance.

The survey also demonstrated that

* Academics support teaching safety in existing courses (60% of respondents) while industrial respondents support teaching safety as a required course (50%).
* Academics spend about 2 to 5% of their time on safety, while industrial respondents indicated spending from 5 to 15% of their time on safety or safety related projects.

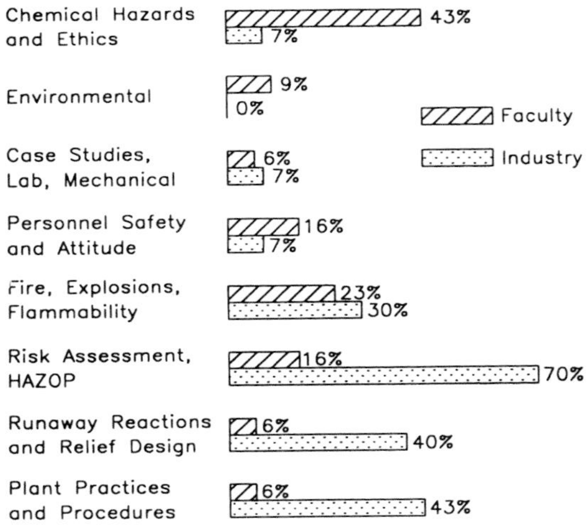

Figure 1:   Faculty and industry response to the question "What safety topics should be taught in universities?"

Engineering accreditation in the United States is managed by the Accreditation Board for Engineering and Technology (ABET). In September of 1986 ABET initiated a study funded by NIOSH to determine the safety and health content of engineering curricula (ref. 2). Each of the 250 ABET accredited engineering schools in the United States were polled for information on the integration of occupational, public and product safety and health in the curriculum. Based on the responses an intensity index was computed. This index was based on course content as well as whether the course was elective or required. Out of the many engineering disciplines, eight were found to rate highest. This included the aeronautical, agricultural, chemical, civil, electrical, industrial, mechanical and mining-metallurgical disciplines. The intensity index for these disciplines is shown in Figure 2. The figure is broken down into safety and occupational health content.

The results show that industrial engineering has the highest rating in safety. This is probably due to the significant attention to human factors. Mining and metallurgical engineering also demonstrated a significant response probably due to required industrial practices in this area.

Figure 2 shows that occupational health ratings in the eight disciplines are quite low. ABET concludes that most engineering programs and most engineering faculty pay more attention to the physical sciences than the life sciences. This leaves the student less prepared to appreciate health issues in the workplace.

ABET also concluded that existing university strengths in health and safety were highly dependent on the initiative of a department head or a department

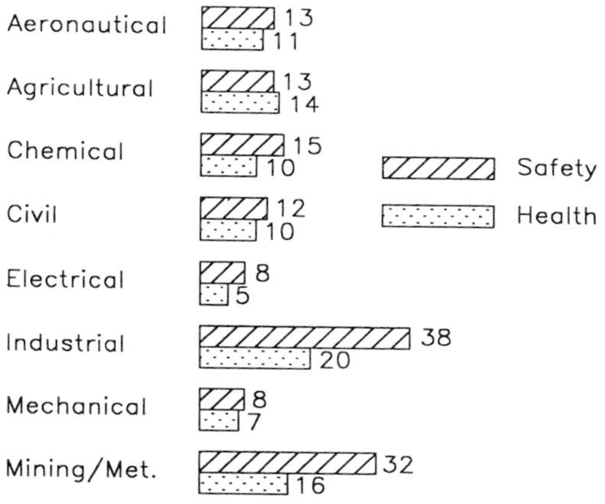

Figure 2:   Safety and health content ratings for eight engineering disciplines.

curriculum committee. It was not dependent on policies implemented by the deans or the upper administration.

At this point in time a substantial dialogue erupted as to the proper mechanism for implementing more safety and health concepts in the engineering curricula. One group argued that these concepts should be added into the existing engineering courses without adding a new required course into the already overcrowded curriculum. They further stated that safety and health is integrated throughout the industrial experience so why not do the same in the academic environment. The opposing group argued that a new required course was the only proper procedure. They stated that, first, integrating the required concepts throughout the curriculum would require the cooperation of all the faculty, a task most difficult to achieve. Second, they felt that many topics in health and safety, such as risk assessment, did not fit properly into existing courses and required a dedicated course.

PROBLEM SET IN CHEMICAL PROCESS SAFETY

In 1986 the Center for Chemical Process Safety (CCPS) of the American Institute of Chemical Engineers (AICHE) formed the Undergraduate Education Subcommittee. The Committee was comprised of industrial and academic members. The charter of the committee was to "identify the key concepts of loss prevention, safety and health which should be considered essential for accreditation of the curriculum by 1990." The Committee identified 34 areas that contain safety, health and loss prevention concepts. These areas are shown in Table 1.

The Center for Chemical Process Safety subsequently issued a Request for Proposal (RFP) to develop of book of problems to illustrate fundamentals of health and safety which could be used in teaching traditional engineering subjects. A contract was ultimately signed with the University of Arkansas under the direction of Professors J. Reed Welker and Charles Springer of the Department of Chemical Engineering. The problem set development was supervised by the Undergraduate Education Subcommittee.

The complete set of ninety problems was published in early 1990 by AICHE under the title Safety, Health, and Loss Prevention in Chemical Processes. An instructor's guide provides additional background material and the problem solutions. The safety, health and loss prevention concept areas were divided into problematic and nonproblematic categories. The problems were developed

Table 1:   The 34 safety and health concepts identified by the
           AICHE/CCPS Undergraduate Curriculum Subcommittee.

1. Properties of Materials
2. Process Design
3. Explosions
4. Toxic Exposure Control,
    Personal   Protective
    Equipment
5. Process Control Interlocks,
    Alarms
6. Toxicology and Industrial
    Hygiene
7. Vapor Releases
8. Inerting and Purging
9. Storing, Handling and
    Transport
10. Fire Protection Systems
11. Hazardous Waste Generation
    and Disposal
12. Rupture Discs and Relief
    Valves
13. Process Hazard Reviews
14. Static Electricity
15. Physical Hazards
16. Human Factors Engineering

17. Regulations, Codes and
    Standards
18. Administrative Decisions
19. Buildings to Resist
    External Pressures
20. Electrical Classifications
21. Sight Glasses
22. Flexible Connections
23. Materials of Construction
24. Equipment Spacing
25. Accident and Near-Miss
    Investigations
26. Hazardous Operations
27. Worker Training and Retraining
28. Types of Process Accidents and
    Relative Risks
29. Preventative Maintenance
30. Concept of Acceptable Risk
31. Emergency Planning
32. Quantitative Risk Assessment
33. Hazardous Material
    Identification
34. Standard Operating Procedures

for use in existing traditional engineering courses such as thermodynamics, fluid mechanics and heat transfer. Nonproblematic concepts were also introduced in the problem description or background material provided the student and instructor.

Subsequent funding by the National Institute for Occupational Safety and Health (NIOSH) and the United States Environmental Protection Agency (EPA) enabled distribution of the student book and instructor's guide to all chemical engineering faculty in the United States free of charge.

The University of Arkansas contributors have received an additional grant from the National Science Foundation to 1) continue development of problem sets, 2) develop and present a number of seminars on chemical process safety and health, and 3) develop and teach a number of short faculty workshops of 2 to 3 days length at various locations across the country to train faculty in safety and health concepts.

VIDEOTAPES ON CHEMICAL PROCESS SAFETY

In late 1986 I received a grant from the National Science Foundation to produce a series of satellite broadcasts on chemical process safety. The broadcasts were to originate live from the pilot plant development area at BASF Corporation in Wyandotte, MI., and were to be broadcast to classroom sites at Wayne State University and Michigan Technological University. The live format would enable students to ask questions during the broadcast.

The motivation for the broadcasts was to learn how live television could be used to bring students together with practicing engineers. Unfortunately, the production difficulties associated with several hours of live broadcast were enormous. First, the demonstrations scheduled for broadcast using actual process equipment would be too difficult to perform. The number of tedious camera moves and close-ups were substantial. Second, the students were familiar with commercial TV, and any decrease in quality would reflect very negatively on the result.

A compromise was achieved by pre-taping as much of the material as possible prior to the live broadcast. The pretaped material would be shown between the live question and answer periods. The live segments would originate in the vicinity of the actual equipment shown in the pretaped segments, enabling questions or clarifications concerning the equipment to be answered. The same BASF engineers presenting the pretaped segments would be available during the live segments to answer the questions directly.

The actual broadcasts were performed during the Fall of 1987. Students responded very favorably to the broadcasts. A post-course evaluation showed that the students felt the broadcast material significantly improved their understanding of chemical process safety.

The project resulted in over 4-hours of pre-taped material covering a wide variety of process safety subjects. The question arose concerning the future of the videotapes. Everyone felt the videotapes could be developed into an excellent instructional resource for industry and universities.

In early 1988, BASF provided additional funding to develop the videotapes into a stand alone product. This project included the following steps: 1) review the existing segments for technical content, 2) re-edit the segments into a more organized fashion for stand alone use, 3) provide introductory and summarizing segments, 4) shoot additional videotape material to expand the content or correct technical problems, and 5) develop a study guide, including a detailed outline and copies of all figures.

The BASF funding was provided under the condition that all revenue could only be used for research and instruction in the area of safety. No funds are to return to BASF or the project team members.

In the Spring of 1988, the existing videotape material was transcribed into a script and VHS copies were made. This was sent to over a dozen reviewers in both industry and academia. In May, June and July a new script was developed. This script included the changes suggested by the reviewers, a substantial reorganization of the material, new material, and introductory and summarizing segments.

In late August and early October, the new material and introductory and summarizing segments were taped at the BASF facility. John Davenport of Industrial Risk Insurers and Stanley Stynes of Wayne State University were the talent for the introductory and summarizing segments.

During October, 1988 through January of 1989 the videotapes were edited and a study guide was developed.

Videotape Content

The main objective of the videotape series is to demonstrate how safety is practiced and how process safety technology is applied in an actual chemical process facility. All of the content was taped on location in the BASF pilot plant facility. While a pilot plant facility is much smaller than a full-sized commercial installation, the safety problems are almost identical.

The material provides excellent content for instruction to 1) undergraduate or graduate chemical engineering students, 2) new industrial employees, 3) industrial employees who have changed job functions, and 4) industrial employees who wish to review safety concepts.

An outline of the final material is shown in Table 2. The segments are organized into 7 different videotapes, enabling the viewer to locate a particular segment easily. No single segment is longer than 30-minutes. The entire package contains approximately 3-1/2 hours of videotape.

Every effort was made to demonstrate key concepts using actual process or laboratory equipment. The video broadcasts were used to illustrate important safety principles which would be difficult to describe in a classroom, especially for students inexperienced with laboratory or plant facilities.

The videotape series comes complete with a 91-page study guide. The study guide is used to 1) take notes during the viewing of the videotapes, 2) studying and reviewing the material after the initial showing, and 3) as a future reference. All of the figures and diagrams shown in the tapes are included in the study guide to assist the viewer.

Table 2:   Outline of Videotape Series: <u>Safety in the Chemical Process Industries</u>

1.   Tape 1, Part 1: Introduction to Videotape Series (4 mins)
     Importance of chemical process safety
     Definitions of safety, health and loss prevention
2.   Tape 1, Part 2: Introduction to Corporate Safety (10 mins)
     Contrasting a good safety program with an outstanding one
     Several ways to achieve an outstanding safety program
     Objectives of a safety program
3.   Tape 2: Laboratory Safety and Inspections (24 mins)
     Safety features and equipment used in the laboratory
     The important parts of a chemical label
     Identifying several hazardous laboratory situations and
      correcting them
4.   Tape 3, Part 1: Process Area Safety Features (17 mins)
     Explosion proof equipment
     Grounding and bonding
     Ventilation
     Vessel overpressure protection systems
     Double block and bleed
     Fail safe concepts
5.   Tape 3, Part 2: Process Area Safety Procedures (14 mins)
     Hot work permits
     Lock-tag-try
     Confined space entry procedures
     Pressure vessel test procedures
     Grounding and bonding during flammable liquid transfer
6.   Tape 4, Part 1: DIERS and VSP (16 mins)
     Problems with runaway reactions and two-phase flow
     Purpose and results of DIERS
     The VSP (Vent Sizing Package) experimental apparatus
7.   Tape 4, Part 2: Dust and Vapor Explosion Apparatus (23 mins)
     Necessity for determining the hazardous properties of
      materials
     Characteristics of dust explosions
     20-liter dust explosion apparatus
     Determination of the Deflagration index
     Blowout panel design
8.   Tape 5: Personal Protective Equipment (19 mins)
     Rules for proper chemical handling
     Specific personal protective equipment demonstrations
9.   Tape 6: Process Area Inspections (15 mins)
     Components of a safety relief system
     Identification of a number of unsafe situations and how they
      are corrected
     Proper procedure for charging hazardous materials
10. Tape 7, Part 1: Informal Safety Reviews (19 mins)
     Re-enactment of an informal safety review
11. Tape 7, Part 2: Introduction to Formal Safety Reviews
     (10 mins)
     The formal safety review report, committee and procedure
12. Tape 7, Part 3: Re-enactment of Formal Safety Review
     (27 mins)

An arrangement was completed in 1988 for the American Institute of Chemical Engineers to market the videotapes. The cost of the complete tape series, including study guide, was established at $500 for universities and $1250 for industry. The low academic price ensured that the material would be available to most universities.

As of the spring of 1990, over 100 copies of the tapes have been sold, including almost 50 universities. This represents a third of all of the chemical engineering departments in the US.

ABET ACCREDITATION CRITERIA

Since 1986, ABET has changed their accreditation requirements to included significant aspects of safety and health. It is important to realize that ABET never specifies actual course requirements but only requires specific course content. Thus it is not possible for ABET to participate in the argument on whether safety and health concepts should be introduced in existing courses or as a new course.

For the 1988–1989 curriculum accreditation requirements, ABET adopted a number of new additions relevant to safety and health. In the section which describes the extent which a program develops the ability to apply pertinent knowledge to the practice of engineering in a professional manner, they require the development "of an understanding of the engineers responsibility to protect both occupational and public health and safety." In the section on engineering design, ABET states it is essential to include a variety of realistic constraints "such as economic factors, safety, reliability, aesthetics, ethics, and social impact." They further state "An understanding of the ethical, social, economic and safety considerations in engineering practice is essential for a successful engineering career. Coursework may be provided for this purpose but as a minimum it should be the responsibility of the engineering faculty to fuse professional concepts into all engineering coursework." Finally, ABET requires that "Instruction in safety procedures must be an integral component of student's laboratory experience."

The AICHE also publishes an accreditation manual which is related in a complicated fashion to the ABET criteria. The 1988–1989 manual states "Although not explicitly required by the criteria at this time the integration of safety, health and loss prevention considerations throughout the chemical engineering curriculum deserves increased emphasis in the view of recent well-publicized chemical process accidents....Not only the technical aspects but the individual, corporate and ethical responsibilities for safety and health should be emphasized in the classroom and the laboratory."

CHEMICAL PROCESS SAFETY: FUNDAMENTALS WITH APPLICATIONS

In 1986 Joe Louvar of BASF Corporation and myself decided that a textbook in chemical process safety was required for university instruction. This would be a textbook in the traditional sense, with an outline which builds on previous concepts, fundamental equations being derived from first principles, worked examples, and homework problems. In fact, the idea was so obvious to us that we were surprised that no one beat us to the finished product.

We decided that an academic/industrial relationship was the proper combination for such an undertaking. The industrial contributor would provide the material relevant to industrial chemical process safety. The academic contributor would insure that the material was presented in a traditional academic textbook fashion.

Furthermore, we wanted to present safety from a quantitative viewpoint, with fundamental calculations being used to make the correct decisions about safety.

A contract was signed with Prentice Hall in November of 1987, the manuscript was submitted in March of 1989, and the book was published in late December of 1989.

An outline of the textbook is shown in Table 3.

FACULTY WORKSHOP

In 1989 I received funding from the National Science Foundation (NSF) to develop and teach a faculty enhancement workshop in chemical process safety, health and loss prevention. The purpose of the workshop was to bring undergraduate chemical engineering faculty together with highly experienced industrial and academic professionals in chemical process safety, health and

Table 3: Outline of textbook <u>Chemical Process</u>
<u>Safety: Fundamentals with Applications</u>

1.   Introduction
2.   Toxicology
3.   Industrial Hygiene
4.   Source Models
5.   Toxic Release and Dispersion Models
6.   Fires and Explosions
7.   Designs to Prevent Fires and Explosions
8.   Introduction to Reliefs
9.   Relief Sizing
10. Hazards Identification
11. Risk Assessment
12. Accident Investigations
13. Case Histories

loss prevention. The instruction was given in a chemical plant environment and demonstrated how fundamental principles are applied in the industrial practice of chemical process safety. Lecture topics included: toxicology, industrial hygiene, source modelling, fires and explosions, relief systems, toxic release, dispersion modelling, and hazards identification and risk assessment. Tours and demonstrations were given using actual process equipment and participants were given "hands-on" experience with a number of process and laboratory safety devices.

A detailed outline of the workshop is provided in Table 4.

The following materials, equipment and special projects were used during the workshop:

1. Lectures on fundamental areas were supplemented by the textbook <u>Chemical Process Safety: Fundamentals with Applications</u> by D. A. Crowl and J. F. Louvar and published by Prentice Hall. Prepublication manuscripts were distributed to all participants. An outline of this textbook is given in Table 3.

2. Lectures were supplemented using the videotape series <u>Safety in the Chemical Process Industries</u>, discussed previously.

3. Tours and demonstrations were given in the BASF Process Development (PD) area. This area is used to develop pilot-scale processes. Safety problems with this equipment are identical to those experienced with full-scale equipment. The equipment included reactors (from 1 to 300 gallons in size), dryers, distillation columns, extractors, carbon absorbers, ion exchange columns, storage tanks, pumps, etc. In relation to the workshop, the BASF facility supported a) eye wash stations, b) safety showers, c) airlines for respirator masks, d) safety reliefs of nearly every variety, e) eXplosion Proof (XP) and non-XP areas, f) grounding and bonding examples, g) extensive personnel protective clothing and devices, h) relief systems, i) on-line monitoring instruments for monitoring safety, including air and water systems, j) deluge systems, k) safety isolation sewers, l) fire and spill alarms, m) inerting systems, n) purge systems for safe drumming procedures, and much more.

4. A half-day tour of the Monsanto chemical plant in nearby Trenton, Michigan was provided. This facility produces organic phosphates and automotive glass products. This tour strengthened the participants understanding by showing how safety is practiced in a somewhat different facility.

5. A complete emergency response drill was performed in the adjacent BASF polymer plant. This drill involved a solvent tank fire scenario and was enacted without the prior knowledge of the plant personnel or the local Wyandotte fire department. The drill demonstrated the use of deluge systems, monitor nozzles, purple-K fire extinguishers, foam systems, and

self-contained breathing apparatus. The drill also demonstrated a number of procedures. This included alarm procedures, evacuation procedures, SCBA recharge procedures, process shutdown procedures, and fire department response procedures.

The workshop was held during the two-week period from August 14 to August 25, 1989. There were a total of 12 participants from universities around the country. The participants were housed in an local hotel and all rooming expenses were paid by the grant. The participant or participant's institution paid for transportation to the workshop and board. BASF provided lunches and a banquet. The cost of all instructional materials was also paid by the grant.

A substantial number of BASF and Monsanto employees and other industrial experts provided instruction during the workshop.

A pre- and post evaluation survey of the participants showed that they agreed that the workshop significantly increased their knowledge and awareness of chemical process safety, health and loss prevention. The survey indicated the participants all felt that the industrial participation was an important component of the experience. Many of the participants indicated that their concept of safety, health and loss prevention was changed by the workshop.

The participants also indicated that the two week workshop was too long. They suggested the period be reduced to less than one week, with a possible extension in lecture time from 7 hours to 10 or more.

Table 4: Faculty Workshop Outline

I. Introduction
    A. Welcome
    B. BLEVE film
    C. Videotape 1, Part 1: Introduction to Videotape Series
    D. Videotape 1, Part 2: Introduction to Corporate Safety

II. Tour of BASF Research and Development Area

III. Toxicology and Industrial Hygiene
    A. Introduction / historical aspects / fundamental concepts
       of industrial hygiene
    B. Industrial toxicology and routes of occupational exposure
    C. Chemical hazards - some examples
    D. Noise / ergonomics / bio-hazards
    E. Film - Hazard Communication
    F. Industrial hygiene evaluation - air sampling
    G. Control methods
    H. Film - Respiratory Protection
    I. Government regulations overview
    J. Industrial hygiene equipment demonstration
    K. Videotape 5: Personnel Protective Equipment

IV. Source Models
    A. Necessity for and usefulness of
    B. Different types
       1. Liquids through holes and pipes
       2. Vapor through holes and pipes
    C. Flashing liquids
    D. Pool evaporation
    E. Examples

V. Probability Theory for Hazards Assessment
    A. Necessity of and usefulness of
    B. Probability distributions
    C. Interactions between process units in series and parallel
    D. Revealed and unrevealed failures
    E. Examples

Table 4: Continued

VI.   Accident Investigations
    A. Learning from accidents
    B. Layered investigations
    C. Investigation summary
    D. Aids to diagnosis
    E. Aids to recommendations

VII. Guidelines for Hazards Evaluations
    A. Introduction
    B. General description and categorization of hazard
       evaluation procedures
    C. Guidelines for selecting hazard evaluation procedures
    D. Checklists
    E. Safety reviews
    F. Relative ranking techniques
    G. Preliminary hazard analysis
    H. Failure modes, effects and criticality methods
    I. Fault trees
    J. Event trees
    K. Cause consequence analysis
    L. Human error analysis

VIII. Introduction to Reliefs
    A. Definitions
    B. Location of reliefs
    C. Relief types
    D. Relief scenarios
    E. Runaway Reactions
    F. Videotape 4, Part 1: DIERS and VSP
    G. DIERS technology
    H. VSP and dust bomb demonstration
    I. Relief installation practices

IX.   Relief sizing
    A. Conventional spring operated reliefs in liquid service
    B. Conventional spring operated reliefs in gas service
    C. Rupture disk reliefs in liquid service
    D. Rupture disk reliefs in gas service
    E. Two-phase flow during runaway reaction reliefs
    F. Venting for fires external to process vessels
    G. Reliefs for thermal expansion of process fluids

X.   Toxic Release and Dispersion Modelling
    A. Necessity for and usefulness of
    B. Fundamental equations and assumptions
    C. Plume model
    D. Puff model
    E. Demonstration of CHARM computer simulation for toxic
       release

XI.   Tour of BASF Process Area to Examine Reliefs

XII. Safety Reviews
    A. Videotape 7, Part 1: Informal Safety Reviews
    B. Videotape 7, Part 2: Introduction to Formal Safety
       Reviews
    C. Discussion of formal and informal safety reviews
       1. What are they?
       2. When are they used?
       3. Who performs them?
       4. Documentation
       5. Implementation
    D. Videotape 7, Part 3: Formal Safety Review

Table 4: Continued

XIII. Plant Safety and Emergency Response Program
    A. Engineering safety reviews for plant installations
    B. Corporate management response program
    C. Plant emergency response drill

XIV. Fires and Explosions
    A. Reasons for concern
    B. Combustion incidents
    C. Nature of combustion
    D. Statistics on industrial accidents
    E. Definitions
        1. Deflagration
        2. Detonation
        3. Autoignition
        4. Others
    F. Basic technology on combustion
    G. Movie on flammable fogs
    H. Oxidants
    I. Flammability limits
    J. Videotape 4, Part 2: Dust and Flammability Apparatus
    K. Venting, containment, suppression
    L. Explosion venting movie
    M. Venting thrust forces
    N. Swiss movie on venting
    O. Problems on venting
    P. Applications of venting
    Q. Weak-seamed roof tanks
    R. Inerting and purging
    S. Static electricity
    T. Other ignition sources

XV. Plant Features and Procedures
    A. Videotape 2: Laboratory Safety and Inspections
    B. Videotape 3, Part 1: Process Area Safety Features
    C. Videotape 3, Part 2: Process Area Safety Procedures
    D. Hot work permits
    E. Confined space entry
    F. Grounding and bonding
    G. Videotape 6: Process Area Inspections
    H. XP electrical equipment
    I. Ventilation
    J. Grounding systems
    K. Control room
    L. Other features

XVI. Tour of Pilot Plant for Demonstration of Key Concepts

XVII. Fire Protection
    A. Fire protection concepts
    B. Tour demonstration fire protection

XVIII. Case Histories

The industrial presenters were not formally surveyed. However, verbal discussions found that they were enthusiastic about the opportunity to provide useful information to faculty and to learn more about the academic experience. In several instances the industrial presenter used the opportunity to "find the time to learn all of the formal details" about a particular subject of interest.

The workshop was highly successful in providing faculty an opportunity to learn about important industrial technology in chemical process safety, health and loss prevention. The industrial presenters also benefitted by providing an opportunity to strengthen their understanding of these important subject areas.

Conversations with the participants in February of 1990 indicated that most of them have incorporated the information learned during the workshop into existing or new courses in chemical process safety. Many of them have also developed important relationships with local chemical companies and are working to develop their own research and instructional programs.

PRESENT STATUS OF SAFETY EDUCATION AT AMERICAN UNIVERSITIES

American universities have improved substantially in safety and health education in engineering. Since 1986 when the problem was first recognized 1) a variety of instructional materials have been produced, including problem sets, textbooks and videotapes, 2) faculty workshops have been developed and offered, and 3) accreditation requirements have been improved.

The quintessential question is "What actual instructional activity is occurring at American universities?" At present I am personally aware of at least fifteen chemical engineering departments (out of some 140 total) which have active instruction in chemical process safety and health. These programs are mostly due to the efforts of a single individual faculty member. Since almost 50 copies of the videotapes have been sold, it can be concluded that these universities have at least some component of instruction (although it might consist of simply showing the videotapes). At present I am not aware of any required courses in health and safety. Michigan Technological University is implementing a required 10-week quarter course beginning in the Fall of 1991. Also, the University of Alabama is working on implementing a required 1-credit semester course and a 3-credit elective. The 3-credit elective can be taken instead of the required course.

Most universities have opted for either of two approaches. In the first approach the safety and health instruction is provided entirely as an elective course. In the second approach, the instruction is provided as part of an existing design or laboratory course. Several universities provide dedicated safety instruction during several weeks of the senior design course. A separate textbook is frequently specified for this segment of the instruction.

The real question is how the ABET accreditation site inspectors interpret course safety content. The ABET criteria for safety and health instruction has not been implemented long enough to determine the impact. However, it is clear that a large number of universities are concerned. It is uncertain whether a single elective course (not taken by all students) will satisfy the requirements. It is also not certain how difficult it will be to demonstrate course content when the content is scattered throughout the curriculum. If this results in difficulty then it will force universities to use more identifiable means such as required courses or dedicated segments in existing required courses.

Recently a new problem has arisen. Chemical engineering enrollments are currently undergoing a modest expansion. A previous large enrollment expansion was followed by a considerable enrollment downturn. As a result, university administrators are reluctant to hire additional faculty lest the enrollments take a downturn again. At present, chemical engineering departments are diverting diminishing faculty resources back to existing courses rather than expanding efforts in new areas, such as safety and health.

SUGGESTIONS FOR IMPLEMENTATION IN OTHER COUNTRIES

The first requirement for implementation is to provide awareness of the problem. This awareness must be present in the faculty and the university administration. The motivation for improvement is quite simple: If university graduates wish to compete with American and European engineers, then additional safety and health instruction is essential.

Adequate class instructional resources are available, but are mostly written in English.

The next requirement is faculty training in these new areas. It is now possible for faculty to reach a certain level of safety and health competence through self instruction. A previously mentioned textbook by Crowl and Louvar (ref. 3), the previously mentioned videotapes marketed by the AICHE, the problem sets available from AICHE/CCPS (ref. 4), a series of instructional modules by IChemE (ref. 5), and a variety of references (ref. 6-28) are suitable for this purpose. Much of the newer technology, such as risk assessment, is either undergoing significant transition or the required reference material is still diffuse.

It is highly recommended that faculty develop relationships with industry. Most of the technology in safety and health was developed by industry; they are the experts and the users of this technology. Faculty must talk to industrial people about safety and health, should spend a summer working at an industrial facility, and should invite industrial experts to offer seminars at the university on safety and health topics. Faculty might also join with industry and faculty at other universities to offer workshops at industrial locations to learn about this technology.

Changes in accreditation requirements are essential to obtain the attention of university administrators.

Faculty will need to be creative in adding safety and health in an already overcrowded curriculum. Some suggestions are 1) a one credit elective course with an elective replacement 3-credit course, 2) a well-defined instructional segment in an existing course, such as design, and 3) integration of the material throughout the entire curriculum. I believe the full 3-credit course is still the easiest and most identifiable technique for adding this material (but it will take some effort to rearrange the curriculum).

Finally, faculty should transmit the correct attitude to students. Safety and health is more than technology; it is really an attitude, just like honesty and integrity. If the students complete a degree with the correct safety and health attitude and a minimal amount of technological tools, then they will make an aggressive effort to use safety properly.

SUMMARY

This paper has presented a detailed historical account of safety and health instruction at engineering colleges in the US and has presented suggestions for implementation at universities in other countries.

The implementation in the United States required a number of important developments:

* The development of essential instructional resources, including problem sets, textbooks and videotapes.
* A change in the accreditation requirements to include more safety and health instruction.
* An improvement in faculty awareness and background in safety and health through faculty training workshops.
* Creative efforts to implement this new material in an already overcrowded curriculum.

The instructional materials at present are more than adequate for classroom use (although they are mostly in English). The last three requirements must be implemented by the faculty in their respective countries.

REFERENCES

1. Louvar, J. F. and Crowl, D. A., "Safety: Opinions of Industry and Universities," to be published in CHEMTECH.

2. Kubias, F. O., "CCPS Undergraduate Education Committee Activities," 1990 Loss Prevention Symposium, San Diego, CA, August, 1990.

3. Crowl, D. A. and Louvar, J. F., <u>Chemical Process Safety: Fundamentals with Applications</u>, Prentice Hall, Englewood Cliffs, NJ, 1990.

4. <u>Safety, Health and Loss Prevention in Chemical Processes</u>, American Institute of Chemical Engineers, New York, NY, 1990.

5. <u>Hazard Workshop Modules</u>, The Institution of Chemical Engineers, 165-171 Railway Terrace, RUGBY, CV21 3HQ, United Kingdom.

References 6 through 28 are recommended for self instruction and/or reference.

6. Lees, F. P., <u>Loss Prevention in the Process Industries</u>, Butterworths, London, 1986.

7. Kletz, T. A., <u>HAZOP and HAZAN, Notes on the Identification and Assessment of Hazards</u>, 2nd ed., The Institution of Chemical Engineers, Rugby, UK, 1986.

8. Kletz, T. A., <u>Cheaper, Safer Plants or Wealth and Safety at Work</u>, The Institution of Chemical Engineers, Rugby, UK, 1985.

9. Kletz, T. A., <u>What Went Wrong? Case Histories of Process Plant Disasters</u>, Gulf Publishing, Houston, TX 1986.

10. Kletz, T. A., <u>Learning from Accidents in Industry</u>, Butterworths, London, 1988.

11. Kletz, T. A., <u>Critical Aspects of Safety and Loss Prevention</u>, Butterworths, London, 1990.

12. Fawcett, H. H. and Wood, W. S., <u>Safety and Accident Prevention in Chemical Operations</u>, 2nd Ed., Wiley-Interscience, New York, 1982.

13. Shrivastava, P., <u>Bhopal: Anatomy of a Crisis</u>, Ballinger Publishing, Cambridge, MA, 1987.

14. Bodurtha, F. T., <u>Industrial Explosion Prevention and Protection</u>, McGraw-Hill, New York, 1980.

15. Marshall, V. C., <u>Major Chemical Hazards</u>, John Wiley, New York, 1987.

16. <u>Industrial Ventilation: A Manual of Recommended Practice</u>, American Conference of Governmental Industrial Hygienists, Cincinnati, OH, 1986.

17. <u>Guidelines for Hazard Evaluation Procedures</u>, American Institute of Chemical Engineers, New York, 1985.

18. <u>Guidelines for Use of Vapor Cloud Dispersion Models</u>, American Institute of Chemical Engineers, New York, 1987.

19. <u>Guidelines for Safe Storage and Handling of High Toxic Hazard Materials</u>, American Institute of Chemical Engineers, New York, 1988.

20. <u>Guidelines for Vapor Release Mitigation</u>, American Institute of Chemical Engineers, New York, 1988.

21. <u>Workbook of Test Cases for Vapor Cloud Dispersion Models</u>, American Institute of Chemical Engineers, New York, 1989.

22. <u>Guidelines for Technical Management of Chemical Process Safety</u>, American Institute of Chemical Engineers, New York, 1989.

23. <u>Guidelines for Chemical Process Quantitative Risk Analysis</u>, American Institute of Chemical Engineers, New York, 1989.

24. <u>Guidelines for Process Equipment Reliability Data</u>, American Institute of Chemical Engineers, New York, 1989.

25. The SFPE Handbook of Fire Protection Engineering, National Fire
     Protection Association, Quincy, MA, 1988.

26. Fire Protection Handbook, 16th ed., National Fire Protection
     Association, Quincy, MA, 1986.

27. Bartknecht, W., Explosions, Springer-Verlag, Berlin, 1981.

28. Olishifski, J. B., Fundamentals of Industrial Hygiene, 2nd ed., National
     Safety Council, New York, 1986.

Safety education at universities

Trevor A. Kletz

Department of Chemical Engineering
University of Technology
Loughborough, Leicestershire LE11 3TU, United Kingdom

Abstract - This paper explains the reasons why process safety and loss prevention should be included in the training of undergraduate chemical engineers and, based on UK experience, discusses subjects for inclusion.

INTRODUCTION

To be able to join the UK Institution of Chemical Engineers as a full (or corporate) member it is necessary to have a degree in chemical engineering from a University or College which follows an approved syllabus and at least three years experience. The syllabus includes safety and loss prevention though they do not have to be taught as a distinct course; they may be taught as part of a design or other course.

The UK practice is unusual. In most countries, including the US, the majority of chemical engineering students get little or no training in loss prevention. Why then do we in the UK think that loss prevention is an essential part of the training of chemical engineers?

WHY INCLUDE LOSS PREVENTION IN UNDERGRADUATE COURSES?

There are several reasons:

1. **Not an add-on extra**
   Loss prevention should not be something added on to a plant after design like a coat of paint (though sometimes it is) but an integral part of design. Hazards should, whenever possible, be removed by a change in design rather than by adding on protective equipment. A chemical engineer need not know much about paint - an expert on paint can tell him which sort to use - but he should not leave the safety of his plant to the safety adviser. If he does so the safety adviser will add on protective equipment of various sorts - trips, alarms, fire protection etc - to control the hazards. Only a chemical engineer can remove the hazards by changes in design.

   An example may make this point clearer. Nitroglycerine (NG) used to be manufactured in a batch reactor containing about a tonne of material. The operator sat in front of the reactor, watching the temperature and controlling the addition rate of the reactants. If the reactor got too hot it would blow up, taking the operator with it. To make sure that he stayed awake he sat on one of the famous one-legged stools.

   If asked to make this process safer, what would most chemical engineers (and most safety engineers) do? They would add onto the reactor instruments for measuring temperature, pressure, flows, rate of temperature rise and so on and then use these measurements to operate valves which stopped flows, increased cooling, opened vents and drains and so on. By the time they had finished the reactor would hardly be visible beneath the added-on protective equipment. However, when the NG engineers were asked to improve the process they did not proceed this way. They asked why the reactor had to contain so much material. The obvious answer is that the reactor is large because the reaction is slow. But the chemical reaction is not slow. Once the molecules come together they react quickly. It is the chemical engineering - the mixing - that is slow. They therefore designed a small well-mixed continuous reactor, holding only about a kilogram of material, which produces about the same output as the batch reactor. The new reactor resembles a laboratory water pump. The flow of acid through it sucks in the glycerine through a side-arm. Very rapid mixing occurs and by the time the mixture leaves the reactor the reaction is complete. Similar changes have been made to the later stages of the plant where the NG is washed and separated (ref. 1).

Similarly, if they are concerned about the large inventory of hazardous material in a distillation column, most safety engineers will think in terms of emergency isolation valves and other protective equipment. A chemical engineer who has been trained in loss prevention will consider reducing the inventory by the use of Higee distillation equipment (ref. 2) or in other ways (ref. 3).

2. <u>Everyone is involved in loss prevention</u>
Most chemical engineers will never use much of the knowledge they acquired as students; they may never have to design a distillation column, for example, or operate a furnace. But all chemical engineers, whether they work in production, in design or in research, will have to take decisions on loss prevention, will have to identify the hazards on a new or existing plant, decide how far to go in removing them and the most appropriate way of removing them, and see that action is taken. Young engineers need to be made aware that they are responsible, both morally and legally, for the protection of their company's employees and of people who live near the plant and that they are the custodians of their company's assets and reputation. Nelson writes "... safety or health responsibilities are often thrust upon them immediately on arrival in industry ... What industry needs is also a major requirement for those going into public service - a balanced viewpoint from which safety and health risks are weighed against the benefit from and cost of specific measures" (ref. 4).

Universities which give no training in loss prevention are not preparing their students for the tasks they will have to undertake once they graduate.

3. <u>Loss prevention teaches principles, not just applications</u>
It is sometimes argued that the job of a university is to train students in the principles of a subject and that applications should come later when the student enters industry. However, accounts of serious accidents of the past can be used to illustrate scientific principles and to show how accidents can be prevented by the application of basic knowledge.

For example, in 1966 at Feyzin in France, a leak occurred from a pressure vessel containing propane under pressure. The leak ignited and a fierce fire burned underneath the vessel. The available water was used to cool neighbouring vessels, to stop the fire spreading, it being assumed that the relief valve would protect the vessel exposed to the fire. After 1 1/2 hours the vessel burst, killing 18 people and injuring 81. The refinery staff had failed to realize that when the upper portion of the vessel, not wetted by the liquid, was heated by the fire it would lose its strength and burst even though the pressure was at or below the set point of the relief valve.

I have discussed the Feyzin fire on many occasions with groups of students and also with groups of engineers from design and production and their reaction is (or used to be) usually the same: Since the vessel burst there must have been something wrong with the relief valve; it must have been too small or not in working order. Only after they have been assured that the relief valve was OK do they realize that the vessel burst because the metal got too hot and lost its strength. They can then work out for themselves that to prevent other vessels bursting in similar circumstances they should be protected by insulation and/or water cooling and/or the pressure in the vessels should be reduced far below the relief valve set pressure.

Everyone knows that metal loses its strength when heated but the staff at Feyzin (and, at first, the people at my discussions) had failed to apply that knowledge. By discussing Feyzin students (and others) can learn the importance of applying basic knowledge and they learn that to say "relief valves prevent vessels bursting" is a myth, a widely held belief that is not wholly true (ref. 5). However, today people realize why the vessel burst much more quickly than they did ten years ago.

Similarly, discussions of explosions of high-boiling point materials such as heavy fuel oils, often considered 'safe', lead to the realization that if they are above their flash points they are as dangerous as petrol.

Finally, not every one would agree with the statement at the beginning of this section, that the job of a university is to train students only in the principles of a subject. In the UK there is long tradition of giving practical training to scientists without compromising on the teaching of fundamentals. Reviewing a book on Victorian chemistry, Hamlin writes, "... The 'pure science' curriculum they developed turned out to match remarkably well the kinds of tasks that industrial chemists ended up doing; ... at the center of the curriculum was a saleable skill; ... Professors argued that chemistry was worthy because it combined practicality with mental discipline" (ref. 6).

WHAT SHOULD BE INCLUDED IN A LOSS PREVENTION SYLLABUS?

The 1983 edition of the Institution of Chemical Engineer's "Scheme for a Degree Course in Chemical Engineering" (ref. 7) listed the subjects which should be included in the syllabus. The 1989 version (ref. 8), in contrast, includes a briefer list of contents followed by an example syllabus. The list of contents merely says that "The teaching of the course must develop an awareness of the necessity for safe design and operation and of social and environmental responsibilities". The example syllabus includes the following under the heading "Operability, reliability and hazard analysis" (no mention of safety or loss prevention):

Systematic identification and quantification of hazards, hazard and operability studies (HAZOP).

Legal background Factory Acts, Health and Safety at Work Act. Legislation on emissions, toxic substances, fire, explosion.

Disposal of solid wastes. Gaseous and liquid emissions, dispersion. Flammability assessment and fire prevention, pressure relief and venting.

Safety of plant in start-up, operation, shut-down, maintenance and modification.

Safety of personnel, and personnel protection systems and procedures.

Reliability and maintainability of equipment. Plant performance and product quality standards, downtime, maintenance frequency.

The Consumer Protection Act 1987, international transport codes and product liability are listed elsewhere in the example syllabus.

Missing from the syllabus is any mention of management for safety (included in the 1983 syllabus), inherently safer design, audits, human error or incident investigation.

In listing a brief general requirement followed by an example the Institution may have been influenced by UK law on safety which now states objectives to be achieved but does not say how they must be achieved, though advice is available in codes of practice and guidance notes.

This section amplifies some of the items in the Institution's example syllabus.

1. Hazard and operability studies (hazops): I suggest that the best way of teaching them (to mature students as well as undergraduates) is to describe the technique briefly (taking, say, half an hour) and then let the students apply it to a line diagram. A suitable diagram for this purpose is shown in reference 9. Ideally the class should not exceed 12 to 15 people but this is not always possible in practice. At least 2 hours are required (including the introductory talk) and the students should realize that they are not expected to get a right answer. There is no right answer. Reference 9 gives the answer that was obtained in the original study but the class may consider that the original team went too far or not far enough and they could be right. If students carry out their design project in groups, they can be asked to hazop at least part of the design.

2. Quantification of Hazards: Systematic quantification of hazards is specifically mentioned in the Institution of Chemical Engineers syllabus and reference 10 is based on lecture notes that I have used for several years. Ideally at least 3 hours are required. Aspects that should be covered are the calculation of probabilities including fault trees, pitfalls, criteria and simple applications such as the estimation of test frequencies. The objective is not merely to introduce students to methods of calculation but to get them to realize that a systematic and numerical approach is possible to questions which at first sight do not seem amenable to quantitative treatment, such as "How far should we go in removing hazards?" and "Is the cost of this measure justified by the size of the risk?"

3. Inherently Safer and User-Friendly Design: Although this is not specifically mentioned in the Institution of Chemical Engineers example syllabus it is such an important part of loss prevention and will become increasingly so during the lifetimes of today's student that it should be included in every course. An hour's lecture should be adequate. Alternatively the concept can be introduced into lectures on other subjects, inherently safer distillation being introduced into lectures on distillation and so on. However, this requires a commitment by the entire teaching staff which may not be forthcoming. Students may also be asked to apply the principles of inherently safer design to their design project. Reference 3 was written as an aid for the teaching of inherently safer design.

4. <u>Human Error</u>: Although not specifically mentioned in the Institution of Chemical Engineers example syllabus human error is an essential part of management for safety, which was included in the 1983 syllabus. Many industrial accidents are said to be due to human error. Students should be encouraged to consider the extent to which accidents can be prevented by persuading people to take more care, the extent to which we should try to change the work situation, that ist, the design or method of operation. One method of encouraging this consideration is by discussion of accidents that have occurred and the action needed to prevent them happening again. References 11 and 12 contain suitable examples and reference 12 also covers accident investigation, another aspect of management of safety.

5. <u>Maintenance and Modification</u>: These are specifically mentioned in the example syllabus and they can supply sets of notes and slides (ref. 13) which can be used for the discussion of accidents that have occurred involving maintenance and modifications (and other subjects). The discussion leader describes the accident very briefly, the class question him to establish the rest of the facts and then say what <u>they think</u> ought to be done to prevent the accident happening again. As with hazop a small group is desirable but not always possible in practice. If students carry out practical work such as dismantling a relief valve or pump they should be required to complete a permit-to-work before removing the equipment from a plant, even if the 'plant' is no more than a few pieces of pipe. A major cause of industrial accidents is errors in the preparation of equipment for maintenance.

A number of publications produced by the Institution of Chemical Engineers as aids for the teaching of loss prevention have been mentioned but anyone teaching the subject should be familiar with the standard work on the subject, <u>Loss Prevention in the Process Industries</u> by F.P. Lees (ref. 14).

A SPECIAL COURSE?

Loss prevention is becoming almost a separate branch of chemical engineering. Is there therefore a need for a special undergraduate course for those who wish to specialize in loss prevention? I do not think so, as the loss prevention specialist requires a good grounding in chemical engineering generally and experience of design or production, preferably both, if he is to be effective and if those experienced in design and production are going to take notice of his advice. They will not listen to an engineer who has taken a course in loss prevention but lacks practical experience. Of course, if a university offers a number of options there is no reason why they should not include one on loss prevention, in addition to the basic loss prevention course taken by all students.

REFERENCES

1. N.A.R. Bell, <u>Institution of Chemical Engineers Symposium Series</u> No. 34, p. 50 (1971).
2. C. Ramshaw, <u>The Chemical Engineer</u> No 389, p. 13 (1983).
3. T.A. Kletz, <u>Cheaper, Safer Plants or Wealth and Safety at Work</u>, Institution of Chemical Engineers, Rugby, UK, 2nd edition, p. 43 (1985). Third edition, entitled <u>Plant Design for Safety - A User-Friendly Approach</u>, will be published by Hemisphere, New York, end 1990. French translation by J.C. Kayser, <u>Des Usines Meilleures Marchées et Plus Sûres ou Prospérité et Sécurité au Travail</u>, Commissariat à l'Énergie Atomique (1985).
4. D.B. Nelson, <u>Plant/Operations Progress</u> 1, No. 2, p. 114 (1982).
5. T.A. Kletz, <u>Improving Chemical Industry Practices - A New Look at Old Myths of the Chemical Industry</u>, Hemisphere, New York, Section 2 (1990). Earlier edition translated into French by J.C. Kayser, <u>Les Mythes de l'Industrie Chimique</u>, Commissariat à l'Énergie Atomique (1988).
6. C. Hamlin, <u>CHOC (Center for History of Chemistry) News</u>, p. 17 (1985).
7. <u>A Scheme for a Degree Course in Chemical Engineering</u>, Institution of Chemical Engineers, Rugby, UK (1983).
8. <u>First Degree Course including Guidelines on Accredition of Degree Courses</u>, Institution of Chemical Engineers, Rugby, UK (1989).
9. T.A. Kletz, <u>Hazop and Hazan - Notes on the Identification and assessment of Hazards</u>, Institution of Chemical Engineers, Rugby, UK, 2nd edition, Chapter 2 (1986).
10. As ref. 9, Chapter 3.
11. T.A. Kletz, <u>An Engineer's View of Human Error</u>, Institution of Chemical Engineers, Rugby, UK, (1985). French translation by J.C. Kayser, <u>Le Point de Vue d'un Ingénieur sur l'Erreur Humaine</u>, Commissariat à l'Énergie Atomique (1988).
12. <u>Hazard Workshop Module: No 008 Human Error</u>, Institution of Chemical Engineers, Rugby, UK (1986).
13. <u>Hazard Workshop Modules: No 2 Hazards of Plant Modifications, No 4 Preparation for Maintenance, No 7 Work Permit Systems</u>, Institution of Chemical Engineers, Rugby, UK, various dates.
14. F.P. Lees, <u>Loss Prevention in the Process Industries</u>, Butterworths, Tonbridge, UK, 2 volumes (1980).

Panel 2: Safety Education in Industry

MODERATORS

S. Cox and G. T. Clegg

Present safety education in the chemical industry of Taiwan

Ada Wen-Shung Ma Lin

Industrial Safety and Health Research Program,
Industrial Technology Research Institute
Hsinchu, Taiwan, R.O.C.

Abstract - In order to launch an effective safety and health enforcement program, the Council of Labor Affairs has requested the Industrial Technology Research Institute (ITRI) to develop a three stage training program to upgrade industrial work environment monitoring and analysis ability. The program consists is as follows (1) To select six core laboratories to assist ITRI in developing 196 analytical methods for each of which the chemicals ROC OSHA has issued PEL level,(2) To develop an interlaboratory comparative testing program, and (3) To develop intensive training courses for sampling and analysis to upgrade the technical capability of industrial analytical personnel. In this one year period some promising results have been shown. The seven "core laboratories" have already vastly improved standards and quality controls. Good participation was obtained in the comparative testing program. The writing of analysis procedures in Chinese is seen to facillitate this process. A summery of the progress is reported.

INTRODUCTION

As with many newly industrialized countries, the economy of the Republic of China (ROC) has consisted largely of many small businesses. Most of these businesses lack both the knowledge of the importance of the safely handling chemicals and of the technology to do so. Furthermore, because most industrial technologies were bought from foreign countries, very little "failure experience" was transfered. Thus, many safety measures have been unwittingly ommitted from various industrial practices in the interest of cutting costs.

A recent ROC OSHA survey (ref.1) indicated 48% of the labor force in the R.O.C. are working in the chemical processing industry. However, over 60% of the inspected factories have been found to lack proper knowledge of chemical management. Another disturbing finding is that the concept of workplace exposure monitoring is new to most industries. 64% of the inspected factories do not performed work environment monitoring; those that do, at best only use detection tubes. Standard methods for air sampling, air contamination analysis, and overall air quality control are unknown.

In an effort to protect workers' health and to prevent occupational disease, ROC OSHA assigned permissible exposure levels (PEL) for 196 chemicals and initiated a self reporting system of work place monitoring. In this way, while these 196 PELs would be ROC OSHA inspection items, routing control would be the responsibility of the individual factory. Unfortunately, in most cases the ability to perform quantitative analyses on the pollutants is seriously lacking.

In order to launch an effective safety and health enforcement program, the Council of Labor Affairs asked the Industrial Technology Research Institute (ITRI) to develop a training program to upgrade industrial work environment monitorting and analysis ability. The program consists of three stages:

1. ITRI selects six laboratories in the country that have sufficient analytical chemistry abilities which will serve, with ITRI as "core laboratories". These laboratories will assist ITRI in developing analytical methods over the next three years for the 196 chemicals for which ROC OSHA has issued PELs.

2. ITRI develops an interlaboratory comparative testing program and invites all interested laboratories to participate in. The program will serve as a preliminary program for accredidation of laboratories able to perform quantitative analyses.

3. ITRI and the six other core laboratories develop intensive training courses for sampling and analysis to upgrade the technical capability of industrial analytical personnel, so that their laboratories may pass future accredidation test.

A summary of the progress of this program follows.

SELECTION AND DEVELOPMENT OF CORE LABORATORIES

As of August 1990, a laboratory selection committee had been formed, the six other core laboratories have been selected, and analytical methods for 40 chemicals have been developed.

Formation of the selection committee

Ten analytical chemistry professors from various universities were asked to serve as committee members. They were responsible for inspecting the candidate laboratories and evaluate their performance.

Laboratory selection

Initially, 41 laboratories were given applications. Most of the university departmental laboratories had been specifically invited, due to their involvement in the fields of environmental science or public health; others laboratories requested applications after hearing about the program through a press conference. However, only fifteen completed the application, and only eleven of those met the preliminary qualifications. The eleven laboratories were asked to take a skills test and were interviewed to see if they fufilled some basic core laboratory requirements. Of the final six laboratories, two each are located in the northern, central, and southern regions of the island of Taiwan. Three are university departmental laboratories, two are nonprofit laboratories, and one belongs to a consulting firm.

Skills testing

To each of the laboratories were sent four organic charcoal samples, each with a different benzene-xylene mixture, and four inorganic spike samples, each with a different lead-cadmium mixture. The laboratories were asked to determine the concentrations of all eight samples within two weeks.

Basic core laboratory requirements

Figure 1 indicates that of the 196 chemicals for which ROC OSHA has assigned PEL's, 75% can be analysed using fully complemented gas chromotography (GC) and atomic absorption (AA) apparatuses. Of this 75%, GC with FID and ECD and AA with flame and graphite furnace set-ups can successfully analyse 90% of these, or 68% of the total.

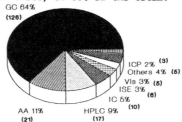

(A) Applicability of general analytical methods for the 196 ROC OSHA-listed chemicals.

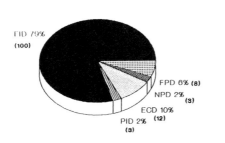

(B) Applicability of different GC techniques to the 126 ROC OSHA-listed chemicals analysable by GC.

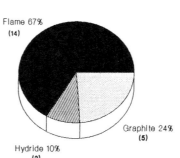

(C) Applicability of different AA techniques to the 21 ROC OSHA-listed chemicals analysable by AA.

Fig. 1. Percent applicability of different analytical methods

Thus, the minimum requirements for the laboratories include :

(1) GC with FID and ECD, and
(2) AA with flam and graphite furnace appartuses.

In addition, to assure the necessary infrastructure for future responsibilities, the laboratories must also have the following :

(3) a laboratory director with published papers, as proof of his research abilities,
(4) sufficient laboratory space for hands-on demonstrations, and
(5) sufficient administrative support and organizational commitment <u>in writing</u> to the mission of the core laboratory.

<u>Selection of chemicals for analytical method development</u>
The list of 196 chemicals were sent to ROC OSHA's eight inspection units and four Council of Labor Affair's field offices for prioritization. Of those, chemicals that could not be analyzed with GC & AA were removed. The resulting list was then discussed among the seven core laboratories to select the final 40 chemicals (Table I) for which analytical methods were developed. The reference methods used came mainly from NIOSH's <u>Analytical Methods, Third Edition</u> (ref. 2).

Table 1. The 40 chemicals for which analytical methods were developed

mercury	nitrobenzene	dimethyl formamide
aniline	nitrotoluene	dichloroethyl ether
calcium	acrylonitrile	cobalt dusts & fume
antimony	dinitrotoluene	copper dusts & mists
pyridine	dinitrobenzene	polychlorobiphenyl (42%)
beryllium	vinyl chloride	1,1-dichloroethane
1-butanol	methyl alcohol	1,2-dichloroethane
2-butanol	isoamyl alcohol	1,2-dichloroethylene
turpentine	tetrahydrofurane	arsenic & its compound
1,4-dioxane	n-Propyl alcohol	cadmium & its compound
iso-Butanol	isobutyl acetate	1,1,1-trichloroethane
naphthalene	trichloroethylene	barium & soluble compound
acetic acid	cellosolve acetate	1,1,2,2-tetrachloroethane
cyclohexanol		

After a method for analysing a paticular chemical was developed by one of the seven core laboratories, explanation of the method would be written in Chinese and the method would be checked by two of the other laboratories for accuracy,reproducibility,and clarity of explanation. Writing the manuals in Chinese turned out to be a particularly successful idea. Many requests for copies of the manuals were received, and the numerous comments about the ease with which the manuals could be used were encouraging. Although there was much initial budgetary opposition to writing the manuals in Chinese, the increasing dissemination is now felt to be well worth the extra cost.

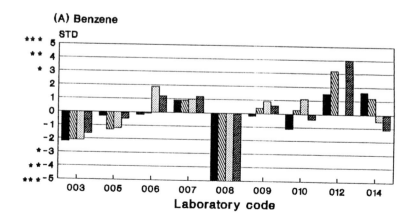

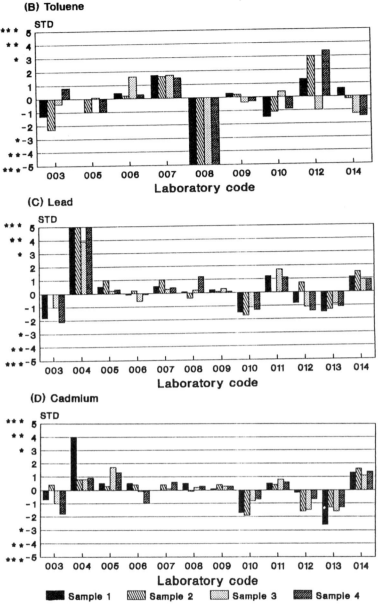

Fig. 2. Results of comparative testing, ROC PAT round 001: organic and inorganic compounds

INTERLABORATORY COMPARATIVE TESTING PROGRAM

An interlaboratory comparative testing program was initiated in December 1989 as a pilot program for a Proficiency Analytical Testing (PAT) program, which will be part of the future laboratory accredation program. Since then, the testing has been conducted quarterly. The purpose of the testing program is to let industrial laboratories know where its abilities stand relative to what would be expected in the future and to what their peers were currently able to do. The results of the first two rounds are given in Figs. 2, 3, and 4.

To evaluate the ability level of the native laboratories, the coefficients of variance (CV) and the failure rates from ROC PAT were compared with those of NIOSH (USA) PAT (ref. 3). Table 2 gives a comparison of CV's for rounds 1 and 2 of ROC PAT with those for rounds 99 and 100 of

NIOSH PAT. As expected, both the organic and inorganic CVs of NIOSH PAT are significantly smaller than those of ROC PAT. Table 3 compares the failure rates of ROC PAT with NIOSH PAT. "Failure" is defined by having more than one result disqualified. Disqualification occurs either by a result being 6% or more away from the true value (an accuracy check), or being more than three standard deviations from the mean of the results of all participants (a precision check). The failure rates for the ROC PATs are much higher than those of NIOSH PATs. It is evident that there is yet much room for improvement in the performance of the ROC's analytical laboratories.

TABLE 2. Comparison of CV values of NIOSH PAT and ROC PAT. 380 laboratories participated in NIOSH PAT; 30 participated in ROC PAT.

	NIOSH PAT				ROC PAT			
	Round 99		Round 100		Round 001		Round 002	
	Inorg	Org	Inorg	Org	Inorg	Org	Inorg	Org
	3.5%	6.4%	4.1%	4.8%	9.4%	6.5%	6.7%	6.4%

TABLE 3. Comparison of failure rates of NIOSH PAT and ROC PAT.

NIOSH PAT *		ROC PAT **	
Inorg	Org	Inorg	Org
9.8%	7.4%	11.1%	22.9%

* Average of rounds 99 and 100.
** Average of rounds 001 and 002.

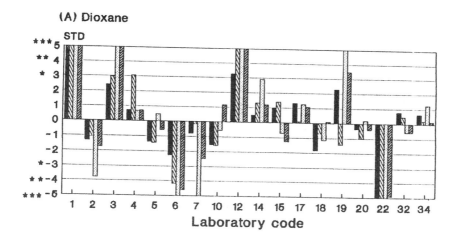

(A) Dioxane

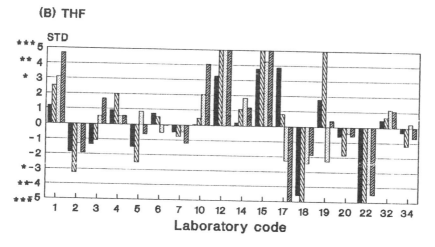

(B) THF

(C) m-Xylene

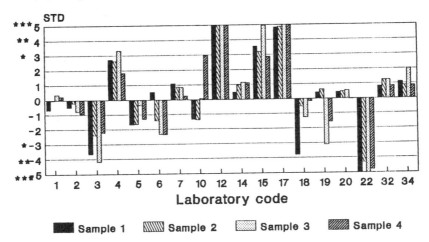

Fig. 3. Results of comparative testing,
ROC PAT round 002: organic compounds

(D) Lead

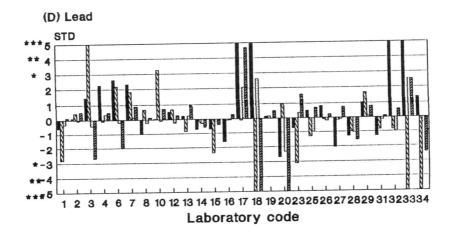

(E) Cadmium

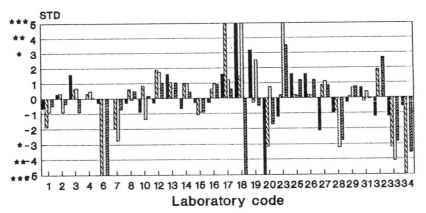

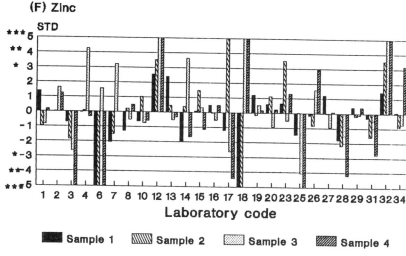

Fig. 4. Results of comparative testing,
ROC PAT round 002: inorganic compounds

CONCLUSION

ITRI's program for upgrading industrial work environment monitoring standards has been well recognized. Response to the core laboratory candidate search has been enthusiastic, and there was good participation in the comparative testing program. The development of Chinese written manuals was also quite useful. Although the current abilities of the ROC's analytical laboratories fall well short of international standards, the upgrading of those laboratories is believed to be an achieveable goal.

REFERENCES

1."Labor Inspection Annual Report: 1988, 1. 1 - 1988, 12. 31"
   Council of Labor Affairs, R. O. C. (1989).
2. P. Eller, NIOSH manual of Analytical Methods, Third Edition.1,
   DHHS (NIOSH) Publication No. 84-100 (1984).
3. S. Schlecht, S. Shulman, J. Groff, Proficiency Analytical Testing Statistical Protocol,
   NIOSH, Revision QAS87-2 (1987).

Safety education in the chemical industry of Malaysia

Teck Thai Lim

Department of Chemistry, Petaling Jaya, Selangor, Malaysia

ABSTRACT – Safety in the Chemical Industry not only consists of proper plant design with safety measures in areas of transportation, production, consumer protection, and protection of the environment and people living near chemical plants, but also involves the important aspects of initial job instructions on safety which includes safety training courses and workshops. In Malaysia, the Factories and Machineries Act 1967, which was enforced in 1970, is concerned with safety, health and welfare of workers in the factories. The Factories and Machineries Department of Malaysia, together with the National Productivity Centre and with the assistance of International Labour Organisation (ILO) has established training programmes to promote occupational safety and health. Generally, smaller chemical plants pay minimum attention to safety. They tend to cut costs by compromising on safety. Leading chemical companies, however, do not wait for the Factories and Machineries Department to give the directive on safe working practices. They set their own targets, have their own safety improvement plans and safety education programmes.

INTRODUCTION

In Malaysia, the Government is concerned that many employees are exposed to danger at workplaces. As an indication of its concern about safety in its industrial programme and its appreciation of the importance of the right safety standards, the Government has appointed a consultant from ILO to make recommendations on the possible set-up of an Institute for Occupational Safety and Health in Malaysia. The purpose of this Institute is to conduct training courses, provide technical services to the industry, collect and disseminate information and to carry out research on safety and health. The Factories and Machinery Department (FMD) of Malaysia has obtained funds under the United Nations Development Programme (UNDP) to finance two projects. The first is to carry out studies on the various aspects of organisation, financing, staffing, equipments and activities for the proposed institute. The other is to establish a training programme to promote occupational safety and health.

The FMD, which is the law enforcement body for occupational safety in Malaysia, conducted this training programme with the National Productivity Centre (NPC) of Malaysia. The Malaysia Society for Industrial Safety (MSIS) has also conducted such training programmes designed to meet the industry's need to train various category of personnel on occupational safety and health.

The FMD has also conducted safety training programmes with the Federation of Malaysian Manufacturers (FMM), Penang Free Trade Zone Safety Council and other organisations. The Malaysian Employers Federation (MEF) also conducts training programmes on their own on "safety management" and "safety and health at place of work" for its members. Last year FMD conducted 28 courses participated by 950 people. This year they expect to conduct 36 courses, and they are in the process of developing courses on the safety in the use of chemicals.

Previously, very few organisations were committed to the safety and health at workplaces. However, in the eighties, especially during the last few

years, there was a change. More employers especially the bigger ones like
the FMM, MEF and bigger chemical industries were paying more attention to
and promoting safety and health. The year 1985 also saw the formation of
the National Advisory Council for Occupational Safety and Health, whose
functions are to advise the Minister of Manpower on the formulation,
implementation and reviewing of a national policy on occupational safety
and health.

HISTORY OF SAFETY LAWS IN MALAYSIA

The first legislation connected with safety at work was enacted at the turn
of the century with the introduction of the Boiler Enactment of 1903. This
was concerned with the safe operation of boilers usually used in tin mines
and dredges. After the 2nd World War, the Machinery Ordinance was
introduced to cover the safe operations not only of boilers, but other
types of machinery as well. This ordinance was replaced by the Factories
and Machinery Act 1967 and was enforced in 1970 as it was found that the
Machinery Ordinance did not cover hazards that affected the health of the
workers.

FACTORIES AND MACHINERY ACT 1967

The Preamble of the Act states :-
"This is an Act to provide for the control of factories, with respect to
matters relating to the safety, health and welfare of persons therein, the
registration and inspection of machinery and for matters connected
therein."

Briefly, some of the salient points of the Act are :-

a.   Safe place of work in relation to floors, working
     levels, platforms, roofs, access to place of work etc.

b.   Protection of bodily injury from explosives,
     inflammable, poisonous or corrosive substances or
     ionising radiations.

c.   Protection against dangerous parts of machinery.

d.   In respect to health, adequate work area, lighting;
     and adequate ventilations to ensure that air in the
     work place is free of all gases, fumes, dust and
     other impurities that may be injurious to health.

e.   Protective clothing and appliance.

f.   A duty on the owner to maintain all safety appliances
     and machinery at all times.

g.   A duty on workers to use appliances for securing
     safety and health, and not interfere and misuse them.

h.   Training and supervision of inexperienced/new workers
     who are employed on any machine or in any process.

REGULATIONS

Under Section 56 of the Factories & Machinery Act 1967, power is conferred
on the Minister of Manpower to make regulations. In pursuant of this
Section of the Act, several Regulations has been enacted :-

i.   Factories and Machinery (Steam Boiler and Unfired
     Pressure Vessels) Regulations 1970.

     This regulation essentially specifies requirements in
     the design, fabrication, installation and
     commissioning of steam boilers and unfired pressure
     vessels. It includes in detail safety devices and
     attachments for the machinery. Even welders are

required to be vetted and tested by the inspector before fabrication or repair work can be carried out.

ii.   Factories and Machinery (Electric Passenger and Goods Lift) Regulations 1970.

Similarly, this regulation specifies in detail the safety devices and attachments for lifts.

iii.   Factories and Machinery (Fencing of Machinery and Safety) Regulations 1970.

This regulation provides details on guarding requirements for prime movers, transmission machinery and driven machinery. Driven machinery includes grinding wheels, machinery using rolls, woodworking machines, mixing machines, power presses, guillotines, conveyors, centrifuges, machine tools and grinding mills. It even provides detail dimensions for abrasive wheel hoods, and guards for mee rollers and crepe rubber horizontal two roll mills. (Mee is a local noodle)

iv.   Factories and Machinery (Safety, Health and Welfare) Regulations 1970.

This regulation specifies further details on the safety, health and welfare provisions given in the main Act.

v.   Factories and Machinery (Person-in-charge) Regulations 1970.

Machinery required to be in charge of persons holding certificates of competency are steam boilers, steam engines, internal combustion engines or dredges. Dependent on the capacity of these equipment, different grades of person-in-charge are specified.

vi.   Factories and Machinery (Certificate of Competency - Examination) Regulations 1970.

This regulation specifies the experience and qualification required for the 11 types of certificates of competency examinations.

vii.   Factories and Machinery (Administration) Regulations 1970.

This regulation specifies the records to be kept by the Chief Inspector, Senior Inspector and State Inspector and fees payable for obtaining such records by the public.

viii. Factories and Machinery (Notification, Certificate of Fitness and Inspection) Regulations 1970.

This regulation specifies the requirement for initial and subsequent periodic inspections for unfired pressure vessels, steam boilers and hoisting machines. Notifiable accidents and industrial diseases are to be reported in a prescribed form.

ix.   Factories and Machinery (Compounding of Offences) Rules 1980.

The Chief Inspector or Deputy Chief Inspector may offer to compound an offence to the person reasonably suspected of having committed the offence.

x.   Factories and Machinery (Compoundable Offences) Regulations 1978.

This regulation specifies all the compoundable offences in the Act.

xi.   Factories and Machinery (Lead) Regulations 1984.

This regulation specifies the permissible exposure limit PEL and an 'action level' of exposure. If the PEL is exceeded, the employer shall have to implement engineering and work practice controls and provide respiratory protection for his employees. If the 'action level' of exposure is exceeded, the employer shall have to carry out exposure monitoring and medical surveillance of employees. The regulation also specifies housekeeping standards, hygiene facilities, practices, and annual training for employees.

xii.   Factories and Machinery (Asbestos Process) Regulations 1986.

This regulation specifies the permissible exposure limit for asbestos dust. The Employer is required to provide exhaust equipment or personal protective equipment to protect employees. The regulation also specifies housekeeping standards, handling and disposal procedures, exposure monitoring and employee training.

xiii.   Factories and Machinery (Building Operation & Works of Engineering Construction) (Safety) Regulations 1986.

This regulation specifies minimum safe working conditions such as safety helmets, eye protection devices, etc and makes it compulsory for construction sites having 50 or more persons to establish safety committees.

xiv.   Factories and Machinery (Mineral Dust) Regulations 1989 (Effective 1.2.89).

This Regulations specifies the permissible exposure limit for mineral dust. It shall apply to all factories in which any mineral process is carried on, and sand blasting process shall not be used in any factory. The Empolyer is required to provide approved personal protective equipment to Employees. It also specifies housekeeping standards, exposure monitoring and employee training.

xv.   Factories and Machinery (Noise Exposure) Regulations 1989 (Effective 1.2.89).

This Regulations specifies the permissible exposure limit for noise at the workplace. The Employer is required to provide approved hearing protection devices and ensure their usage. The Employees are required to wear these devices provided, co-operate during exposure monitoring, undergo audiometric testing or medical examination and attend training programmes.

The present law and regulations on safety and health, like in most other countries, are based on the traditional regulatory approach; whereby it is believed possible to control and overcome all hazards by means of detailed regulations to be enforced by Government agencies. The system encourages too much reliance on Government regulations and too little on personal responsibility and voluntary self- generating efforts. The attitudes, capacities and performance of people and the efficiency of the organisation within which they work are equally important factors.

It can be seen that the Factories & Machinery Act 1967 has brought about regulations on a piece meal basis to counter new hazards and yet does not cover workers in agriculture, plantation, forestry and transport etc. Besides, its enforcement role may relax management's accountability in new disciplines to safety until such time as FMD comes out with new regulations to counter new hazards.

It is envisaged that Malaysia's own Occupational Safety and Health Act which is in the process of being enacted in the near future will address this short fall. It will incorporate the philosophy that primary responsibility for doing something about the present levels of occupational accidents and disease lie with those who create the risks and those who work with them.

SAFETY IN THE CHEMICAL INDUSTRIES

Malaysia's Health & Safety Act which is scheduled to be launched requires that the employer provide safety training and information as is necessary to secure the health and safety at work of his employees. It is generally accepted that training is an important aspect in a successful safety programme as it increases awareness and changes attitudes.

With Malaysia poised to be an industrialised nation with its output from the manufacturing sector taking the lead again last year, commitment to safety and health should be commensurable to the growth of the manufacturing sector.

In fact, it does! Many chemical industries, especially the bigger ones and those with substantial foreign investments, adopt safety standards which are substantially higher than the minimum standards currently laid down by the Government.

In Malaysia, most of these chemical companies comment that the cost of educating staff in safety procedures is considerable. Even the cost involved in the supply of basic safety wear to employees such as shoes, hats, spectacles, etc. is itself not insignificant. Most bigger chemical industries employ a full-time safety officer as well as consultants, and some organisation's safety procedures are reviewed by their international consultants.

Many of these industries indicate their commitment to safety by means of visible involvement in safety matters by the Chairman or Chief Executive Officer. In fact, most bigger chemical industries have a clearly defined written policy on safety.

HEALTH AND SAFETY POLICY

With management's commitment to safety, workers are aware of the weighting safety gets at their work site. This commitment comes in the form of a Safety and Health Policy, a copy of which is issued to each and every member of the work force when he joins the company.

A typical Safety and Health Policy (from ICI Group in Asean) will read as follows :-

ICI GROUP IN ASEAN

HEALTH AND SAFETY POLICY

It is the policy of the ICI Group in Asean that in the manufacture and use of its products, proper care will be taken to protect the safety and health of all persons involved.

It is the policy of the company to manage its activities so as to avoid causing unnecessary or unacceptable risk to the safety and health of employees and customers and of any members of the public who may be affected by its operations. This policy means that the company will conform with legal requirements and appropriate codes of practice

and will take any additional measures it considers
necessary. The company will make its knowledge and
expertise available to the relevant authorities.

Most of the safety policies are based on the following :-

i.   That safety of employees, the public, and company operations are
      paramount.

ii.   That safety will take precedence over expediency or short cuts.

iii. That every attempt will be made to reduce the possibility of
      accident occurrence.

The policies of several of the companies, that I have come across, seem to
be clear, communicated ones that show visible management commitment and
full acceptance of line responsibility and accountability. Even the safety
of contractors are assigned to those who employ them. They will have to
ensure that the contractors comply with all aspects of works safety
regulations. They will have to ensure that the contractors' personnel
clearly understand the particular hazards of working on their chemical
plants, for instance, and, when necessary, the personal protection and
hygine precautions required.

The established policies are rigidly enforced and made effective by a safe
system of working.

SAFETY ORGANISATION

It is generally recognised that competent safety advisors are required to
primarily advise, train and monitor safety activities. Great importance is
attached to the resolution of safety problems as close to the place of work
as possible.

From the several chemical firms surveyed, there is usually a Safety
President heading the safety organisation which comprises the Chairman of
the Safety Committees of the various Plants/Sections. These chairmen are
usually the Section Managers or Plant Managers and each committee is
represented by a cross-section of the staff of the Plant/Section, including
the Engineering staff.

In the Chemical Company of Malaysia Berhad Shah Alam Works, the various
safety committees meets once in 2 months to discuss and resolve safety
matters and do inspections of their workplace.

SAFETY TRAINING AND EDUCATION

Safety in the Chemical Industry not only consists of proper plant design
with safety measures in areas of transportation, production, consumer
protection, and protection of the environment and people living near
chemical plants, but also involves the important aspects of initial job
instructions on safety which includes safety training courses and
workshops.

Most chemical industries in Malaysia provide some basic safety training and
education, as well as specific safety training. The small industries do it
on an ad-hoc basis and training is usually done "on-the-job". The bigger
industries do have a training program as shown by Tiram Kimia's (a Shell
associate company) example :-

	Description	By	When	Remarks
1.	Fire/ Evacuation drill	Safety President/ Chairman, Operations Supt	Quarterly	For all employees

2.	Fire-fighting using hydrants	Safety committee	December	For Fire-fighting teams
3.	Fire-fighting using portable extinguishers	Safety committee	December	For office staff
4.	Safe Forklift driving	Local dealer	December	For those who drive forklifts
5.	First-Aid	First-Aiders	June	For all employees
6.	Respiratory equipment, Self-contained breathing apparatus, Gas mask	Production & Operations Supt	June	For plant and operations personnel
7.	Safe operating procedures	Production & Operations Supt	December	For plant and operations personnel

At the Shah Alam Works of the Chemical Company of Malaysia (ICI group), all plant staff are given 6 monthly training on basic first-aid, basic fire-fighting and use of breathing apparatus and other personal safety equipment.

Qualified (Red Crescent, the local equivalent to the Red Cross) first-aiders are trained and made available in every shift. They are given regular refresher courses by the Safety Technician to maintain their certificates.

A fire-fighting team is maintained in every shift. Regular fire-fighting practices and training are conducted for every team by the Site Supervisor, with the help of Officers from the local Fire Department.

All new plant personel are given anti-gas training. They are given safety induction training as outlined in their Safety Induction For New Recruits manual.

Safety training on methods of carrying out job safely, including process and maintenance operations, are conducted by their respective Plant Managers/Engineers.

At the Sterling Drug company, new personnel are subjected to basic fire-fighting training, which is conducted once a month by the company's fire teams and the Fire Department. Specific training on fire-fighting is also carried out by sending the staff to the Fire Department's training school and this is done twice a year. Fire drills are performed once a month and all personnel are trained on basic first-aid, emergency procedures etc.

At Pacific Chemicals (of the Dow Chemical Company Group), safety education is provided from the time he/she is just hired. The company ensures that the new employee is made aware of the possible hazards and must be responsible for his or her own, as well as co-workers' safety.

The following items are emphasized during their safety indoctrination program of each new full-time and temporary employees :-

   1.   Organization of department and department safety program.

2.   Detailed description of hazards existing in work area.

3.   Restrictions on clothing and footwear.

4.   Eye protection and respiratory protection.

5.   Smoking regulations.

6.   Procedures for disposal of waste materials.

7.   Procedures for handling dangerous materials and equipment, and for storing dangerous materials.

8.   How to handle minor injuries, policy concerning trips to the medical facility.

9.   Location and use of emergency equipment (shower, fire blanket, extinguishers, fire hose, gas masks, etc).

10.   Location of emergency exits.

11.   Recognition of alarm signals, and evacuation procedures.

12.   Location and use of safety manuals and other literature.

SAFETY AUDITS

Safety Audits were conceived when management realised that there was no upper limit for safety improvements. Being familiar with ones plant area one may overlook some of its pit falls, or on the other hand one may be reluctant to highlight unsafe conditions lest this should give rise to more work. It may, therefore, necessitate the calling of an expert from an external body to perform a Safety Audit.

A typical Safety Audit from a chemical plant in Malaysia involves :-

	WHAT	BY WHOM	WHEN
1.	Department/Section Self-audit	Dept/Section Supervisors	Quarterly
2.	Plant audit	Safety committee	Twice a year
3.	Unsafe Act Audit	All	Randomly
4.	Safety standards audit	Safety committee	Once every two months
5.	Safety/Housekeeping inspection	Safety committee	Monthly
6.	Occupational Health audit	Shell Medical Advisor	Annually

Safety committees play an active role in improving work place safety. Their participation in safety audits will go to improve safety standards in work places and stimulate involvement and commitment of both the management and workers. The phychological effect of workers playing a part in decision making are likely to result in a more productive work force.

A typical Safety Audit programme practiced by the Safety Committee of a local chemical plant is as follows :-

February   1.   Safety Standards audit :

	a. Loading & Unloading Bulk Lorries
	b. Safe Work Permit Procedures
	c. Hot Work Permit Procedures
	d. Line and Equipment Opening Permit

April
1. Safety Standards audit :
   a. Leak and Spill Control
   b. Eye Protection
   c. Contractor Safety

June
1. Safety Standards audit :
   a. Office Safety
   b. Fire-fighting
   c. Emergency Planning

2. Plant audit

August
1. Safety Standards audit :
   a. Means of Egress
   b. Lock-out of Power-driven Equipment
   c. Testing & Emergency Alarms and
      Protective Devices

October
1. Safety Standards audit :
   a. Hoists
   b. Compressed Gas Cylinders
   c. Personal Protection

December
1. Safety Standards audit:
   a. Hoses
   b. Vessel Entry
   c. Electrical Lock-out and Red Tag
      Procedures

2. Plant audit

CONCLUSION

The management/worker federation for safe working in Malaysia has ushered in an era for safety training, monitoring and campaigns all necessary to induce changes in human attitudes thereby leading him to manage safety positively, reduce potential for accidents and thereby minimise the risk of major industrial accidents occurring.

We should achieve maturity age in terms of safety with the importation of technology, which also pulls along safety standards. Malaysia is not far behind compared with the developed countries. The safety concern is mainly on the cottage-type industries.

Generally, smaller chemical plants pay minimum attention to safety. The employees might feel it is inconvenient to wear personal safety devices. Employers might try to make do with the simplest safety equipment to keep a certain profit level. However, leading chemical companies do not wait for the enforcement agencies to give the directive on safe working practices. They set their own targets, have their own safety improvement plans and safety education programmes.

REFERENCES

1. Factories and Machinery Act, 1967 and Its Regulations.

2. Abdul Jalil Mahmud (1987). Overview report of Occupational Safety and Health in Malaysia.

3. Hamed Samad (1983). Review of Safety in the Industrial Laboratory.

Safety education in the Japanese chemical industry

Yoichi Uehara

2

2

2

2

2

2

2

2

2

2

2

2

2

2

2

2

2

Department of Safety Engineering, Faculty of Engineering, Yokohama National University, 156 Tokiwadai, Hodogaya-ku Yokohama 240, JAPAN

ABSTRACT-Education in industry is closely related with the school education system. A young man who is a graduate from a high school and obtained employment in the chemical industry is sent to the workplace of the plant after being educated about fundamental knowledge on company and chemical engineering. At this stage, he is again provided with on-the-job training and starts a career as a full-fledged technician of the plant in several years. In this paper his records of attainment together with affairs of safety education for university graduates, text books, several practices utilizing simulator, and KYT (Kiken Yochi Training: Hazard Prediction Training) are described. Thus, the reason why few accidents occur in Japan is examined.

INTRODUCTION

Chemical industry provides many useful materials to other industries and homes. However, since large amounts of dangerous substances are sometimes handled in the plants or factories, once an accident breaks out, the extension of its damage is incomparable to that of other industries. It goes without saying, therefore, that the safest conceivable precautions should be taken at every step of the industrial preparation operations such as the selecting of materials, process or structural materials, or in planning and constructing the process and plants etc. Also, safe measures should be taken by improving instruments/equipment and installing safety equipment. Since all these things are handled by people, industrial plants remain inoperative without their knowledge and actions. It is indispensable, therefore, to provide them with appropriate knowledge and safety education. This report will familiarize you with the present state of affairs in the safety education of the Japanese chemical industry and refer to the problems we will be facing hereafter.

SAFETY LEVELS OF THE JAPANESE CHEMICAL INDUSTRY

The statistics regarding fire and explosion accidents that have occurred in plants and facilities which handle dangerous substances in Japan are shown in Fig. 1. This figure shows the numbers of facilities handling dangerous materials and of cases of accidents and accident occurrence rate drawn from these numbers. As Fig. 1 clearly shows, while the number of the facilities has been increasing, that of cases of accidents have been either leveling off or decreasing. Consequently, the accident occurrence rates have been gradually declining and have dropped to about the half in the past 20 years. The accident occurrence rate at type A facilities which produce dangerous articles out of dangerous materials is 10^{-3}/facility annually while that at type B facilities which merely handle dangerous substances is 10^{-4}/facility annually. These rate are considerably better than the fire occurrence rate at office buildings and residential houses which stands at 10^{-3}/house annually, in consideration of the difference in the circumstances. This excellent improvement seems to be attributable to the technological advances, tightening of legal controls and a sufficient training in safety education. Causes of accidents occurring at facilities handling dangerous articles lie partly in hardware type of factors such as defective planning or deterioration of facilities/plants and partly due to human factors such as managerial deficiencies or operational errors.

Looking more closely into the factors, it is clear that lack of appropriate
knowledge contributes largely to the causes of these accidents.   While

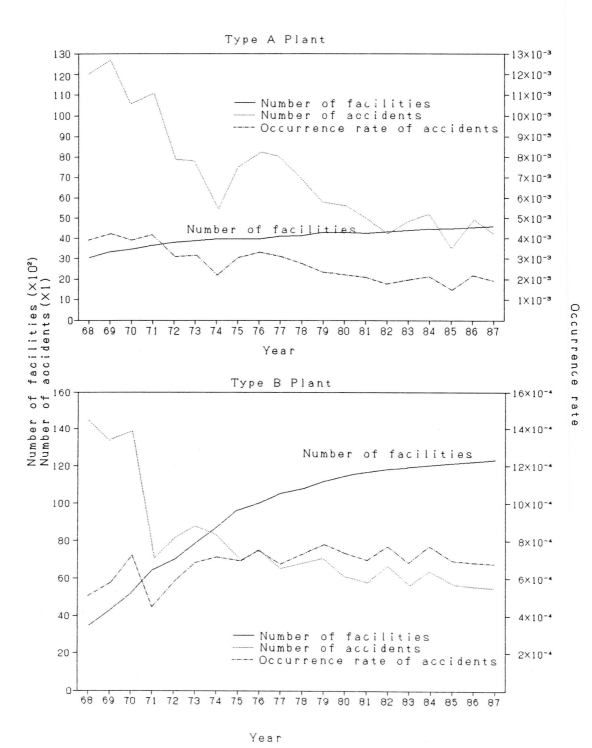

Figure 1   Number of Cases and Occurrence Rate of Fire and Explosion
         Accidents at Facilities Handling Dagerous Substances.

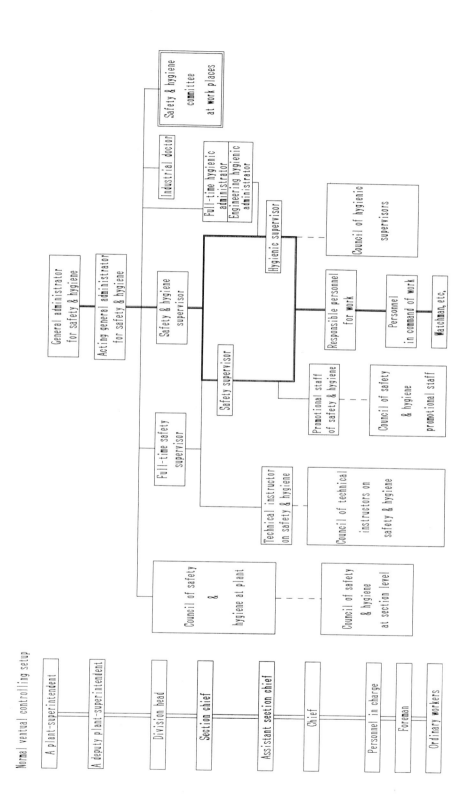

Figure 2 Safety and Sanitation Controlling Structure at a Plant

accidents caused by hardware are relatively easy to control, those induced
by human factors are difficult to do so. Hence, the importance of employee
education given by companies has been significantly increased.

ORGANIZATION IN PLANT

In a plant, the production department which naturally takes charge of
producing chemical substances has the largest workforce. It is divided and
subdivided according to the types of the chemical substances it produces.
The division which is in charge of the technical problems of the safety and
environment at a plant is the environment and safety division while a safety
and hygienic section mainly takes care of the health of employees and of
accidents involving their injuries or death. The engineering division is
mainly in charge of repairs and maintenance of facilities and plants.

In the case of a plant on a scale of having a workforce of about 1,500,
roughly half (about 750) of the workforce belong to the production
department, of which roughly 500 are the shift workers.   The plant has
approximately 200 university graduates, the half of which are engaged in
research/development activities while the other half work either at the
production sites or in the administrative department.

Whereas the above is a vertical management system, there is a separate
common organization on safety and hygiene with lateral contact in each
sections. This organization which is shown in Fig.2 deals with safety and
hygiene as common problems of whole workplaces and has a right to order to
improve their situations, if necessary.

This paper will argue mainly on educational issue for operators and
technicians who work in a production department in a plant,with matters of
relating departments to be commented as required.

BASIC POLICY ON EDUCATION

The purpose of educational in Japanese companies is to increase the ability
of individual workers for the benefit of the company as well as to broaden
the scope of the individual worker's life. Therefore, the education covers
a wide range of subjects, ranging from knowledge necessary for a worker to
perform his or her duty to various lectures on cultural courses.   A
comprehensive ability development program is outlined in Fig. 3 while a
curriculum of training conducted at each work site is shown in Table 1.

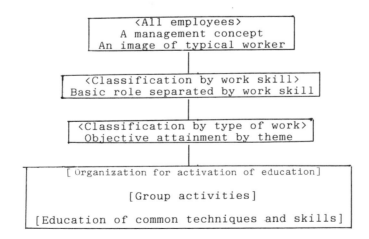

Figure 3   Schematics of Ability Development System

Principle objectives of safety education are to conduct stable operations
without accidents or damages, to keep employees in good health and to
eliminate ill-effects on neighboring communities.

Table 1   Educational System of a Company

	Mandatory education	
	Name of education & training	How they are conducted
Managing staff	Administrators' club	Lecture meetings
	Education on personnel & labor management	Collective training
Administrative specialty position	Enlightenment training for sectional staff (TQC course)	Lodging-in training and followup education
	Enlightenment training for sectional staff (IE course)	Collective training and followup education
	Ability development training on troubleshooting (KT method)	Collective training and followup education
	Sectional staff adviser training	Collective training
Supervising position	Training to reform foremen awareness	Lodging-in training
	Enlightenment training for foremen (TQC course)	Collective training and followup education
	Enlightenment training for foremen (IE course)	Collective training and followup education
	Foremen candidate training	
Specific to light-work position	Specific work 4D training	Collective training
	TPM education	
	Training upon LL·LS appointments	Collective training
	Group leader training	Collective training
		Visit of other industries, participation in exhibitions
	Training on business manner	Participation in the course
	Young workers training	Collective training
	New employees training	Collective training

EMPLOYMENT AND EDUCATION OF WORKERS

After World War II, the Japanese school system underwent a drastic reform. In the reformed system, education starts with children of 6 years of age, who undergo to 6 years of elementary school education, followed by 3 years each of junior high-school and senior high-school education. A basic Japanese education system is completed with 4 years of college/university education. In a different vein of education, there are higher technical schools for junior high school graduates. While compulsory education ends with completion of the junior high schools, the ratio of students who go on to senior high school currently amounts to 95%. While in the past, those who had completed the compulsory education were taken on as operators, the trend has been changing, with senior high school graduates taking up the positions of operators. And now that the ratio of senior high school graduates who go on to colleges/universities is almost reaching 50%. Then graduates of higher technical schools also sometimes work as operators.

Employees who work in a plant are divided into two groups according to the types of work they are engaged in; one is a group of skilled workers such as operators and the other is a group of technical workers. The former group are engaged in operating plants while the latter group are engaged either in research and development activities in laboratories or in steering and supervising the running of the plant. Normally, graduates of senior high schools are employed as operators and college/university graduates or those with master's degrees are employed as technical staff. Current trend is that graduates of higher technical schools are increasingly taken on as candidates for the staff in charge of work fronts.

Regarding education, general education in a company is handled by the personal administration department while specific technical education is given by senior technical staff with expertise in each specific field as instructors. Regarding safety education, safety and hygienic education is given by safety and hygienic supervisors while environment and security education is handled by the environment and security department. There are subjects of education which should be commonly provided by companies and which should better be given at plants. The way to arrange this varies according to each company.

Senior high school graduates, after receiving the basic education on their companies, are assigned to work sites in plants, half the day taken up by lectures and half by practical training. By the time they get to be able to write flow sheets of the plants they are charged with, in other words, approximately one month later, they are put on three-shift duty. After about 5 months of apprenticeship, they become normal shift workers. However, since they are still, at this stage, inexperienced in their jobs, they work under the guidance of foremen. After working this way for approximately one to two years, with particular importance attached to their familiarization with periodical repairs and startup and shutdown procedures, they become full-fledged workers.

Some companies educate graduates of higher technical schools in the same way as with senior high school graduates while some conduct a collective education for a period of 6 months, training them, in addition to the nature and setup of the companies, on such subjects as chemistry, chemical engineering, machinery, electricity, electronics or biological engineering according to the type of jobs they are to take up. After completion of the training, they are assigned to each work post and receive on-the-job training.

Meanwhile, college/university graduates receive collective education at company head-office for one to three months. Afterwards, half of them are assigned to the research/development department while the other half are assigned to their work posts in the plants. The later will, after going a through apprenticeship on three-shift duty, formally take up the three-shift work as technical staff under the guidance of foremen. In two to three years, they will become conversant regarding the technical workings of the plants. The education of employees consists of off-the-job training and on-the-job training. Off-the-job training includes at-the-desk education and on-the-spot training which is conducted with actual machines as teaching materials sometimes actually running them. An example of a training program using actual machines is shown in Table 2.

It is possible for skilled staff with a senior high schools or higher technical school background to be promoted to the position of a foreman section chief or even head of a division depending on their abilities.

Relationship between level of skill required on the jobs and length of experience is shown in Fig. 4.

In general, the level of skill is raised with the length of experience but some workers are slow to grow in the level of their skill. To qualify to be a foreman, an employee has to be, in general, over 25 years of age with at least 5 consecutive years of the work experience. He or she also needs to pass a test (which consists of language ability test, written test to see scientific and mathematical aptitudes and general knowledge and an oral test) and a recommendation from the head of work section he or she belongs to. Those who have passed the examination become foreman candidates and will receive intensive training, which is to educate them on such subjects as personal management, safety education, trouble-shooting or quality control, etc. Some companies provide a 4 day and 3 night training course which is repeated 4 times. After completion of the training course, those failing to achieve a good mark, are unable to qualify as a foreman. There are some companies which provide training, spanning a period of 2 years, with the remaining curriculum comprising of a collective training, on-the-spot practical training, educational course, on-the-job training and research activities on specific theme, etc.

There are many cases where foreman candidates are requested to obtain licenses to be able to handle dangerous substances or high-pressure gas as well as public qualifications regarding safety and hygiene.

Table 2 An Educational Program with Actual Machines

Subjects of education	Substance of education	Teaching materials & equipment
Security & safety	Experience of various dangerous factors Learn the feeling of fear Principle Knowwhy and prevention Study on avoiding measures	Experimental equipment for particle explosion, static electricity, thermal expansion, water force, liquid seal and ejecta effect
Rotation machinery	Structures, functions, characteristics, trouble anticipation, prior maintenance, abnormality response, and operational method regarding such machinery as pumps, fans, etc.	Actual equipment using electrical power Cut products Experience plant CAI
Electricity	Sequence and actual operations Structures of electrical equipment, their roles, functions, the operational method and judgement on their trouble and response.	Dummy panels Explanatory panels A false load Experience plant CAI
Instrumentation	Instrumentation basic loop and PDI adjustment. Setup and traits of instrumentation system and abnormality response	Analogue 4 loops CRT control plant Cut products CAI
Metal materials	Usage standard on metal materials Principles of corrosion and anti-corrosion Prediction and prevention of corrosion	Various metal materials Specimens of corroded metal material Corrosion experiment equipment
Non-break test	Understand through experience the fact that specific danger lies in welded parts	Various non-break tests training equipment

Even after being appointed as foremen, they will receive higher level training on the same subject matter. In such training for foremen, who are the lowest level management officials, considerable emphasis is laid not only on technical expertise handle subordinates and on the way good human relationship in work place should be . Education continues with even full-fledged skilled workers. Every month, more than 5-hours of education is spent on learning basic things, up-to-date technical information, new methods of safety analysis and control or doing small-group activities, amounting to a total of 100 - 150 hours annually.

If such educational activities are conducted on off-time hours, overtime payment will be duly paid. Staff with college/university education are provided with higher level training which covers such subjects as technique for FTA (Fault Tree Analysis for Process Failure) and FEI (Fire and Explosion Index of Dow Chemicals).

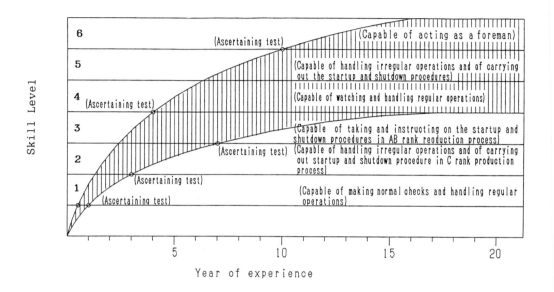

Frontal Work: Level 1 Elementary, 2 Intermediate, 3 advance
Panel Responsibility: Level 4 Elementary, 5 Intermediate, 6 advance

Figure 4   Relation between the Length of Experice and Required Level of Work Skill.

ROLLS OPERATORS ARE EXPECTED TO PLAY

Nature of education and training provided to operators varies according to what they are required of.

Roles operators are expected to play fall into the following four categories:

(1) Control of process operations
(2) Control of plants
(3) Security and safety control
(4) Production control

Control of process operations
Various changes can occur during the process of operations. To counter such changes, operators must always try to grasp the kinetic behavior of the process precisely. Expressed another way, they must try daily to develop their abilities to respond properly to various situations. Since a mistake in responding to abnormal trouble may lead to a major accident and damage, they have to acquire sufficient technical expertise to determine the cause of the abnormality and take corrective steps as quickly as possible. They also have to train themselves to recognize beforehand what kind of situations may develop into dangerous ones and be able to operate the process with confidence, anticipating what results will come about from the steps they are taking.

Control of plants
Although repair, improvement and diagnosis of plants are handled by the maintenance department, responsibility for controlling plants normally lies with the production department which operates the plants. Consequently, operators are the front-line inspectors. Operators have to manage plants in such a way as to prolong the service life of the plants by making early detection of trouble and taking immediate corrective measures if it occurs.

Security & safety control and production control
Since the control of process operations and that of plants entail these two items, it is necessary for operators to have sufficient basic knowledge about them. Operators are also required to have technical expertise regarding improvements in the plants. At present, various improvement activities such as plant/equipment improvement movement conducted with the proposal system and small-group activities, operational technique improvement or security and safety standard improvement, etc. are conducted at each work place in Japan, regardless of the type of industry. These improvement activities are closely related to the performance of the control of process operations and of plants.

Substance of Education

As preliminaries to entering into this subject, you will be familiarized with the practices and what are widely conducted in Japanese industries. Recently, Japanese industrial products have acquired a reputation for excellent quality, the fact of which is attributable to quality control techniques introduced just after the War. Successful dissemination of the concepts and technique owes much to the fact that every worker at every Japanese company has participated in the TQC (Total Quality Control) movement, without leaving the technique to quality control experts alone to promote and implement. These days, another movement called the TPM (Total Productive Maintenance) movement, which is to execute preservation and maintenance by all workers, being is actively undertaken. These movements contribute as much to improving the quality of products as to the safety at work places. Hazard Prediction Training (KYK; Kiken Yoti Kunren or KYT; KY training) is also rigorously being undertaken. The purpose of this training is to train workers so that they will be able to recognize potential dangers that are latent in their work places, to eliminate them and to use countermeasures if they become a reality. Recently, a chemical process version of the Hazard Prediction Training, which is called PKYK (Process KYK), has emerged.

Basic education on safety, although not referred to here, is naturally taught at the early stage of at-the-desk lectures or on-the-job training. Knowledge on chemical substances and on process which is necessary for a worker to work at his or her work post is also given. It is a practice of a Japanese worker to repeat aloud an instruction given by his superior or his intention, which is executed without fail. When starting an action, a worker points at the object of his action with his finger and says aloud the action he is taking. This practice is called "Shisa-kosho" or "point and call." For example, when shutting valve A, a worker goes to valve A and says, "I'm going to shut down valve A" while pointing at it with his finger before taking the action. When the valve has been shut down, he again declares "I've just shut down valve A", pointing at it. This practice goes a long way toward reducing errors stemming from human factors.

The quickest way to understand the substance of education is to see the text. The texts vary greatly with each company. Some texts, which are produced by companies themselves, are good enough to pass current outside of the companies while some others are merely handwritten.

Contents of one of the best texts are listed below. The text consists of three books, with the second and third books having a separate-volume supplement each.

-Heinrich's Law
-Accidents induced by actions and process accidents
-How accidents happen
-Rationality and irrationality
-Combination of safety measures
Chapter 4   Safety education
-Safety education according to law
-Safety education system at our company
Supplement   Guidance to basic behavior

Book Two: Development of Safety Consciousness
Chapter 1   Traits of human beings
-State of human being through cerebrophysiology
-A story of great carelessness
Chapter 2   Elimination of blind spots
-Study on men's traits
Chapter 3   Blind spots which lead to misjudgments
Chapter 4   Blind spots lead to mistaken action
Chapter 5   Study on instances of accident
A separate-volume supplement
A collection of blind spot instances

Book Three: Technique to review safety
Chapter 1   Viewpoint to review safety
Chapter 2   Hypotheses of accident
-Hypotheses of accident
--Simulation of accidents by hypothesizing
--Operability study
Chapter 3   Improvement on safety measures
Chapter 4   Assessment of safety standard
--Assessment of safety efforts
--Diagnosis of safety
A separate-volume supplement
Instances of implementation at a model plant

As previously mentioned, technical staff with college/university education are provided with higher level safety education.

EDUCATION WITH SIMULATORS

Introduction of simulators

The chemical industry original was an equipment-oriented industry, with plants having long been controlled with analogue instrumentation. Recently, however, digital control system with CRT displays are rapidly replacing them.

This technological innovation, coupled with improvements in plant maintenance, contributes to operational stability, thus reducing the chance of operators encountering actual abnormalities or trouble, learning from experience. Decreased though the abnormalities and trouble are, they are still not yet totally eliminated and there is some concern as to the ability of current operators in responding properly to the abnormalities and trouble when they happen, specifically if control system fails. To counter this new situation, it is necessary to adopt a new educational method to complements the area where work experience and the on-the-job training can not cover. It is the reason why the education with simulators has been introduced.

Models

It is possible to use standard models and actual plant models for the educational purpose. The former is unit process models which are commonly used in actual plants such as flash-drums, boilers, compressors or distillation towers while the latter models are actual plants which are used to train the workers in the inherent operations of the plants.

To render the training efficacious, the simulators need to be able to simulate the following functions:
    1) Regular and irregular operations
    2) Changes in the operational conditions
    3) A function to simulate various trouble
    4) A function to increase and decrease simulation speed

5) Playback and replay function
6) Evaluation of the training results
7) Automatic training function

Substance of education

There are many cases where, in consideration of differences in experience, four different levels of educational courses: introductory course, elementary course, intermediate course and advance course, are provided to cover all levels of operators, ranging from new employees to foremen. Additionally, there are some cases where teams of operators, are provided with education while on the duty.

Education is provided to a team of 4 workers by one to two instructors. The team consists of a leader, 2 CRT operators and a field operator and they take turns playing each role as required. The field operator, while in contract with a control room, mans field equipment, which is displayed on the CRT with the actual sound of the machine running, providing verisimilitude. As an example, an educational program, which is a 6 day course, for a monomer plant is shown in Table 3.

Table 3   Training Curriculum for Monomer Model

period	Day	Training menu
The first term	1	- Orientation CRT operation Information gathering activities with CRT Understanding of model process
	2	- Procedure on regular operations Ascertaining of operational conditions Information gathering on operations Changing of conditions - Trial operation of plants Trial operation of equipment Measuring the capabilities of equipment & instruments - Work involving plant operations Preparations of the prestage of startup procedure Execution of regurar operations
	3	- Taking the startup procedure Examination of the startup procedure and production of the manual Execution of the startup procedure - Review of the satrtup procedure Verification and modification of the manual on the startup procedure Reproduction of the startup procedure - Plant test Examination of mass-balance and heat balance Examination of test operations
The second term	4	- Shutdown procedure Examination and execution of the shutdown procedure Simulation of misoperation - Review on the shutdown procedure Trouble with equipment during the shutdown procedure - Corrective steps to be taken in case of irregular operations Emergency shutdown procedure Assessment of danger level
	5～6	- Response training to abnormalities Training to predict occurrence process of an abnormality (mainly abnormality experience) Training with an emphasis on the early corrective measure to counter an abnormality which is easy to detect (mainly efforts to inquire into the cause of an abnormality) Training with look importance attached to the response to an abnormality, the cause of which is difficult to trace (combination of remedical measures for abnormalities and inquiry into the causes)

CONCLUSION - THE WAY SAFETY EDUCATION SHOULD BE IN THE FUTURE

Some of what are described in the foregoing paragraphs may be commonly done world-wide while some may be practiced only in Japan. Each country has its own way of educating its workers. I do not mean to say that our way is the best. It is a pleasure if you know our way as a reference.

Up to now, Japan has been known as a country with a life-time employment system. A Japanese worker, once employed by a company, would stay with the company until the retirement. Japanese education for workers is built mostly on the basis of this practice. As a result, it has taken as long as 10 years to develop a full-fledged operator, which has posed no serious drawbacks. Those who have been underpinning an excellent industrial performance were workers with either compulsory or senior high school education of a period previously. With diligence and aspiration, they have been endeavoring for the sake of their companies and many of them are still active on managing positions. The things, however, are changing drastically. Now the ratio of senior high school graduates who go on to university reaching nearly 50%. It is becoming difficult to rely too much on those workers with mandatory or senior high school education, the stratum of workers who once played a major role in conducting safety operations. The hope is now laid on graduates of higher technical schools to play the part. The major theme for the chemical industry hereinafter is who to employ and how to educate and train them. This theme will directly influence the way future educational system for worker will be shaped.

The Japanese lifetime employment system is also gradually going out-of-date, with the number of cases where workers switch companies increasing. It is one of the major problems that how the new employees with different job backgrounds can be best educated so that they will be able to attain the same level of competence as of those who have been working for the company from the start.

Another problem is that young people tend to show reluctance toward going on three shift duty. It is important to make work places attractive for youth generation to recruit young workers.

These days, an opinion is heard that section chiefs, heads of divisions, directors or even the highest level of managing officials should also be educated on safety as it has little effect if only operators and low-level technical staff are given the safety education. Moreover, some are critical that the present safety education is unbalanced as emphasis is laid partially on technological and technical aspects of the safety, without paying much consideration to a need of social safety or addressing legal aspect of it.

The times are changing rapidly, entailing changes in people and technology. Something good now may not be good forever. But then, something good may be good regardless of the time. As such, safety education in the chemical industry should be rectified as required by the time while addressing the fundamental points regardless of changes with times.

In closing, I would like to express my sincere gratitude to those who have assisted me in preparing this report. Although I will not name any names, much is owed to the people of chemical, oil, petrochemical and pharmaceutical companies who have willingly let me interview them, sometimes providing me with very precious material. Also, it should be noted that model companies appearing in this report were created wholly out of the author's imagination on the basis of various data and that responsibility for all the facts and articles in this report is the author's.

Safety education and training in industry

Sue Cox

Centre for Extension Studies, Loughborough University of Technology,
Loughborough, Leicestershire, UK.

Abstract - This paper considers safety education and training in relation
to the **chemical** industry. It first looks at the context to such
training in terms of organisations' technological, economic, societal
and legislative environments, and then discusses the process of training
in terms of the 'training cycle'. It concludes by presenting a case
study which not only illustrates the points raised but also provides
an example of the author's work at Loughborough University of Technology.

INTRODUCTION

Safety education and training in industry are complex topics. A consideration of, at
least, two issues is necessary to unravel that complexity in a practically useful way.
First, there must be an analysis of the dynamic interactions which occur between
organisations and their technological, economic, societal and legislative environments
(Cox, 1987).   These interactions not only shape the behaviour of those organisations
but also that of the people who work within them.   Second, this analysis needs to be
set in the context of an understanding of the other processes which underpin effective
education and training, and which have been variously referred to as the 'training cycle'
(Cox, 1988a).   This paper briefly considers safety education and training in the **chemical
industry** in these terms.   In doing so, it outlines, through a case study, one of the
programmes of education and training in health and safety currently being provided by
the author through the Centre for Extension Studies at Loughborough University of
Technology.

Technological environment

The complexity of the technology employed by the chemical industry is continuing to increase
and is becoming both more "intelligent" and less dependent on the human operator.   While
the possibility of overall systems failure may have decreased through improved engineering
controls and technological "fixes", the impact of any such failure may have increased
because of the size and nature of plant.   Where operators are still required to play
a role in process control, increased automation may be contributing to operator stress
and to the likelihood of human error (Bainbridge, 1987).   Furthermore, the organisation
and management of such work may be becoming more critical and by their very nature may
give rise to, what Reason (1989) has termed, latent errors: disasters waiting to occur.
Somewhat paradoxically, therefore, one implication of such technological change is a need
to review the training of both process operatives and their managers.

Economic environment

During the period 1970 to 1980, production in the UK chemical industry grew by 3 to 4%.
By 1986, however, the effects of the recent recessionary period had become obvious.
Notwithstanding this, the UK message from 1987 onwards was:

> "a good growth forecast for chemical industry"
> (Annual Business Conference, 1987)

Generally, in the UK, investment in education and training follows the economic cycle,
although often decreasing faster than other markers and recovering more slowly.   Sadly,
the effects of periods of low investment in education and training can feed forward
creating skills shortages in subsequent periods of growth.

Societal environment

In recent years there has been a marked decline in the public esteem of the chemical
industry.   Jack Gow (Secretary General, UK Royal Society of Chemistry), in his Ivan

Levinstein Memorial Lecture (1988), quoted two sources of information.   He drew attention
to a then recent public opinion survey in USA which rated the chemical industry only just
above the tobacco industry in popularity, and also to a Chemical Industries Association
survey in UK which reported a 50% favourable response to the chemical industry in 1979
decreasing to 33.3% in 1988.   In the same meeting, the President, CIBA Geigy (Canada)
is quoted as saying that:

> "chemical corporations now have to operate in an
> environment polluted by fear......."

Legislative environment

Most industrialized countries have developed national rules and laws governing health
and safety at work which are expressed in terms which can be verified by visiting inspectors.
The UK framework is provided by the Health & Safety at Work Act, 1974, (HASAWA), which
was the initial **enabling** Act, and by its associated regulations of which the Control of
Substances Hazardous to Health regulations (COSHH, 1988) and the Pressure Vessel
regulations (1990) are important recent additions.   The UK Health & Safety Executive
(HSE) has the necessary enforcement powers to support the Act and its associated
regulations.

Under this legislation, employees and employers have duties to identify and control hazards
and minimize risks.   One of their general duties is ensure the "education and training"
of their workforce in relation to health and safety (section 2, HASAWA).

The UK Act and its associated regulations do not aim for absolute safety, rather they
seek **a reduction in risk as far as is reasonably practicable.**   Safety is to be secured
by the 'best practicable means'.   The concept of "reasonably practicable" is important
in the UK legislation, and has been defined by case law.   Somewhat in contrast, other
European countries have enacted legislation which attempts to go further than that of
the UK and which sets out to establish absolute safety against quantitative standards
(and the necessary supporting procedures).   Despite this apparently tougher stance in
some European countries, Frank Lees and John Withers (Withers, 1987) have found no evidence,
from reported accident statistics, to suggest any marked differences in the outcome of
safety practices, in fixed chemical process installations, among industrially developed
countries.   Notwithstanding this question of the relative effectiveness of different
legislative strategies, there is a move towards the harmonization of European legislation.
This is now an important issue in the UK and other European countries.

The differences in the legislative environments, within which organisations have to operate,
do have to be taken into account especially by multinational chemical companies which
operate in many different national contexts.   Such companies have a duty to comply, but
problems relating to absolute standards may arise if the relevant national health and
safety laws are not well developed?   Many of the countries in which they operate are
part of the developing (and under developed) world.   Do those multinational companies
therefore need to develop their own 'internal legal systems' set against agreed internal
standards?

There are relevant Codes of Practice in the UK for UK based multinational chemical companies:
they are provided by the Health & Safety Executive, the Chemical Industries Association
and the various relevant professional bodies, such as the Royal Society for Chemistry
and the Institute of Chemical Engineering.   They also reflect the principles of responsible
care.

Education and training are implicit and sometimes explicitly stated in most of the relevant
Codes of Practice.

Organisational response to the challenge of safety

In many ways the dynamic interactions between organisations and these different environments
determine the challenge of safety.   Most organisations have responded to this challenge
in at least one of two fairly limited ways: first by reviewing their hardware and safety
technology, and second by developing their safety software systems including their policy
statements, safety organisation and performance reviews.   Much effort has gone into the
development of common standards and auditing systems, particularly in multinational
organisations.   However, notwithstanding these actions, it has been forcefully argued
that there is a need to carefully consider the human factor and to deliberately develop,
what might be termed, safety liveware.   Where this need is recognised, it is often given
substance through safety education and training.

The development of safety systems, and the promotion of safety education and training,
reflect and in turn are reflected in organisational culture, analysable in terms of the

attitudes and knowledge existing within the organisation, its resources for safety (including its ability to monitor safety performance), its power to influence, and its formal and informal structures.

Safety education and training have become key strands in many accident prevention and safety promotion strategies within UK organisations (Dawson et al, 1988). However it is questionable whether the quality of such education and training programmes has been sufficiently high to produce the expected (and required) return. This is largely due to the failure to properly address all the different processes which underpin education and training and which make up the training cycle (Cox, 1988a). Hale (1984), in a review of safety training, highlighted the fact that most general reviews on this subject (for example, Hale and Hale, 1972) begin their discussions with the comment that remarkably few studies have been published which evaluate the effectiveness of safety training. When questioned about safety education and training, many managers immediately focus on the actual training session; however, this is but one part of an overall training cycle, which begins with the training needs analysis and culminates in the evaluation of training (Cox, 1988a).

SAFETY EDUCATION AND TRAINING

Safety education and training normally seek to fulfill two linked objectives. First, they attempt to improve individuals' awareness, knowledge, attitudes and skills in relation to safety and reflected in safe working behaviour and procedures. Second, they also attempt to effect positive changes at the level of the organisation itself. These include an obvious improvement in overall safety performance. Both objectives are derived (in detail) from the training needs analysis and must be reflected in the later evaluation of training.

Training needs analysis

Logically the training cycle begins with the systematic identification of training needs achieved through a **training needs analysis.** Among other things, this analysis must consider the needs of the organisation in relation to the different environments in which it has to operate (see above).

The training needs analysis should be designed to consider each working group within the organisation in relation to its safety behaviour and to the organisation's overall safety performance. In the chemical industry, these groups may include:

- safety professionals,
- plant process operatives,
- instrument articifers and technicians
- craftsmen and fitters,
- engineers,
- maintenance personnel,
- supervisors and plant management, and
- contractors.

In each case, a task analysis should be completed and safety relevant elements of the task identified and along with associated safe behaviours (Bamber, 1983).

This sort of analysis should allow shortfalls in safety awareness and knowledge, attitudes, skills and workplace behaviour to be identified, group by group, along with priority groups for receiving education and training. By the same process, priority issues or hazards can be identified, group by group, and at the level of the whole organisation or even industry.

A questionnaire based survey by Sandra Dawson and her colleagues at Imperial College, London, on safety in the chemical industry (Dawson et al, 1988), asked respondents (production managers, supervisors and engineering maintenance personnel to identify:

> "the most important health and safety problem or hazard faced
> by the people for whom they (managers or supervisors) were
> responsible or whom (safety representatives) they represented".

Their results are set out in the table below:

Nature of Hazard	% Respondents
Chemicals (toxicity, corrosive, fire, explosion)	37
People (attitudes, carelessness, ignorance)	29
Plant, machinery and systems of work	17
Place of work	12
Other	4

Interestingly, the results demonstrated a relatively high level of awareness, within the UK chemical industry, of issues relating to safety liveware - the human factor. They also pointed out the need to be particularly alert to the needs of several 'at risk' sub groups: new employees, transferred employees, temporary employees, and promoted employees.

Models of safety

This sort of data often allows not only individual and organisational needs to be identified, but also for models of safety behaviour, and of safety systems management, to be developed. These can then act as 'frameworks' for the design and provision of education and training, possibly localized to particular industries, companies, groups or issues. Several examples exist in the general literature: for example, those of Bird (1987) and Cox (1987).

Detailed objectives

Detailed objectives for education and training are developed from the training needs analysis, possibly guided by a 'framework' model of safety behaviour and its management. Not only does the publication of those objectives allow trainees to understand what that training is about, they also effectively guide the rest of the training cycle. The essential question is:

> What should trainees be able to do at the conclusion of training, and how will this differ from what they can do before they are trained?

Objectives should be stated in **behavioural** terms for the benefit both of trainees and of later evaluation. Often this presents terminological and conceptual difficulties when training is targeted on knowledge, attitudes or attitude change (Cox, 1988b). Such difficulties can be overcome with some imagination.

Translation of objectives into practice

Knowing what needs to be achieved through safety education and training is not the same as knowing how it is to be best achieved particularly if such education and training has to be provided for several different groups within one organisation or for different organisations and in different countries. It is argued **here** that those involved in such safety education and training require two things:

- first, a culture free knowledge base, relating to models of safety, basic information on specific areas of safety and good practice (ethics etc), and

- second, the means of tailoring that knowledge base to local requirements and contexts (heuristics or rules for translating generic knowledge into specific scenarios).

This holds true not only when moving from organisation to organisation within an industry or country, but even more so when moving across industries and countries. This can be related back to what has already been said in relation to multi national chemical companies and their response to the differences in legal systems across countries. They tend to establish their own internal code (knowledge base) and develop heuristics or rules for adapting it country by country.

Training method

What is the best way of increasing trainees' awareness, knowledge, attitudes and skills (AKAS) to allow them (enable) them to meet the behavioural objectives set?

Decisions on appropriate training methods will be dependent on a number of criteria (Cox, 1988a), including:

- the nature of the subject matter,
- learning objectives (AKAS),
- the number of trainees,
- trainee preferences and prejudices,
- the total training resource, and
- the convenience factors such as shift patterns and job cover.

Economies of scale are provided by a large number of trainees attending a lecture but better retention of information occurs in more participative scenarios, including role plays. At the same time, computer based training methods allow greater flexibility for individuals within organisations and make allowance for differing rates of learning.

Training materials

Training materials are available from a variety of sources to support safety training and important technological advances have been made in this area with the advent of video discs. Similar developments in computer software (cf: customized permit to work packages) have standardized systems information. Institutions such as the Institution of Chemical Engineers produce training packages on a number of safety related topics including "emergency preparedness". Distance learning materials are also available to industry often prepared in conjunction with educational establishments.

Evaluation of training

How do we know that education and training have worked? Programmes can be evaluated in different ways:

- first, by examining trainees' progress during the course and their reactions to it at the end of the course - internal evaluation,

- second, by examining the impact of the training on the trainee's later job performance (etc) - external validation, and

- third, by asking whether the training has achieved its organisational objectives?

Some computer based training packages build in not only an examination of trainee achievement against programme objectives, but also map their progress in doing so (internal evaluation).

Evaluation needs to be carefully planned into the training cycle, with the design of the evaluation beginning during the training needs and organizational analysis.

EDUCATION AND TRAINING IN HEALTH & SAFETY AT LOUGHBOROUGH

The Health & Safety Management group in the Centre for Extension Studies at Loughborough University currently offers one of the UKs foremost post graduate/post experience Diploma courses in Health & Safety Management. The initial development of this Diploma course was supported by ICI but has now broadened its scope to accommodate a wider range of organisations and safety interests.

The work of the group can be usefully illustrated by the following case study. This particular case is orientated towards the training of managers and other staff who are not, by nature of their work, safety professionals.

Case study

The case study is based on a programme of safety education and training developed for a large multinational company with several different sites within the UK.

The Accident Prevention Advisory Unit, HSE (APAU) had carried out an indepth analysis of these sites, which included:

- bulk chemical and pharmaceutical manufacturing,
- research and
- primary distribution and repackaging of small batch chemicals.

It made a number of recommendations for improving safety performance, including training for middle and senior managers.

Following a training needs and organisational analysis based on three sources of information:

- detailed discussions with the Technical Director, and his safety professionals,

- 'inspection' visits to the larger sites and semi structured interviews with a wide range of personnel within the projected training group, and

- access to the detailed safety audit conducted by the APAU

The author set out a strategy for safety education and training.   This involved:

1. developing general health & safety awareness of relevant areas of H&S law, safety technology, occupational health & hygiene, and environmental pollution control,

2. establishing health & safety as an important line management function, and

3. developing knowledge of specific safety management techniques

in a manner which would enable the company's employees to relate this increased knowledge to their own environments.

Because it was not possible to train all the company's UK employees, a **cascade** strategy was adopted, whereby key staff in all areas would receive training and would be encouraged to pass on their new knowledge to their colleagues.

Training objectives

It was agreed that after training "employees should be able to actively manage safe behaviour and promote safe working within their own working areas".

Trainee population

The specified training population was 4 - 500 managers at all levels, based in factories, laboratories and offices across the various sites.   The majority of the trainees were graduates, aged between 30 and 50 (or non graduates with equivalent professional qualifications).   Most of the trainees were based in the UK, although several managers from oversea's facilities were included in the training group.

Design of training

The training programme was designed to be modular, and encompassed:

- Legislation relevant to company's site operations, including H&S and environmental legislation, role of enforcing authorities, Common Law liabilities (etc),

- Management of safety: roles, policies and organisational arrangements,

- Human factors and the management of people,

- Safe systems of work and working procedures,

- Identification and perception of risk,

- Hazard analysis and assessment techniques,

- Major hazard awareness,

- Monitoring and auditing

-   Environmental pollution and waste management

-   Occupational health and

-   Occupational hygiene.

Delivery and structure of training

The training was based on a series of short courses presented by expert trainers.   These comprised of an initial two day workshop, followed by a work based project, and then a one day feedback workshop.

The 11 modules were presented in these three days (with project): each module was covered in about 2.5 hours by briefings and case discussions, and question and answer sessions, supplemented by company appropriate video and film material.

The programme was supported by the publication of an extensive (management and) training manual.   Each trainee was encouraged to familiarise themselves with the content of this manual so that they might easily use it for future reference.

Evaluation of training

Several different sets of evaluation data are being collected at three different levels. These include:

-   **Immediate & individual** assessments based on post course assessment forms. It is acknowledged that these provide an inherently weak form of assessment, but they generally indicated that the course was favourably received.   Over 75% of the trainees felt that they were more aware and better informed as a result of the course, while a similar percentage felt they had developed a useful understanding of specific safety techniques.

-   **Longer term & individual** assessments based on job performance and achievement monitoring through the company's annual appraisal scheme.   These data allow some comparison of changes in awareness, attitudes and behaviour between those trained and those not trained.   Early indications are that those trained have shown greater safety awareness and commitment, and more detailed knowledge, particularly in relation to the law and specific safety management techniques.

-   **Organisational:** assessments based on a scrutiny of the company's accident statistics (hitherto stable) over the period of the course and onwards.   Early trends suggest an improvement of 30%.

While each of these data is, in itself, circumstantial, taking them together reduces the probability that the positive changes observed are all due to chance.   The value to the company has been demonstrated not only in words, but also by the renewal of the contract.

In reviewing the programme, the author noticed an interesting effect of trainee 'selection' on evaluation.   Although all of the target population (key staff in each facility) had to undergo training, the early courses were populated by those who were most interested and who had been enthusiastic to volunteer.   The final courses were largely made up of "pressed men".   This was more than obvious in two respects:

>   - first, in trainees reactions during the training courses: the
>   "pressed men" were less co-operative and more cynical, and

>   - second, their immediate post course assessments were more negative.

This phenomenon is not something specific to this company, or even to safety training, but simply reflects the interplay of trainee motivation with the impact of training. In a sense what is really needed to overcome this phenomenon is a measure of 'added value' which can take out these initial differences.   This demands at least a pre-post evaluation paradigm.

Other information

Over the courses the training team, themselves all experienced safety professionals, built up a detailed knowledge of the company's safety culture, specific problems and likely future challenges.   This information has been fed back to the organisation through a series of management meetings and has allowed the design of further courses on more specific topics, such as human factors of plant control rooms.

FINAL SUMMARY

This paper has briefly considered safety education and training from various standpoints, and has effectively described a number of overlapping models to guide its development. Six main points have been covered.

- The organisation's behaviour with regard to safety is shaped by its interactions with its technological, societal, economic and legislative environments.

- Education and training should be a common aspect of organisations' accident prevention programmes and response to the challenge of safety,

- Such education and training has to be treated as part of the overall training cycle, which must begin with a training needs analysis which takes into account organisational as well as individual needs,

- The training needs analysis may provide sufficient data for a framework model of safety behaviour to be derived and used in the subsequent design of training,

- To provide training, trainers need a core knowledge base (culture free) and a set of heuristics (rules) to tailor them to local requirements,

- Training must be evaluated and different models of evaluation exist to guide such activities at both the individual and organisational levels,

These points have been emphasised, and in doing so reference has been made to the work being carried out at Loughborough through the presentation of a case study.

REFERENCES

1. Bainbridge, L, In: Rasmussen, J., Duncan, K., and Leplat, J., (eds) New Technology and Human Error, Wiley, Chichester (1987).

2. Bamber, L., In: Ridley, J., (ed) Safety at Work, Butterworths, London.

3. Bird, F.E., and Germain, G.L., Practical Loss Control Leadership, Institute Publishing, Loganville, Georgia (1987).

4. Cox, S., Work & Stress, 1, 67-71 (1987).

5. Cox, S., Health & Safety Officer's Handbook. Vol. 6, 9-15 (1988a).

6. Cox, S., Employee Attitudes to Safety, Unpublished thesis, University of Nottingham, Nottingham (1988b).

7. Dawson, S., Willman, P., Clinton, A., and Bamford, M., Safety at Work: The Limits of Self Regulation, Cambridge University Press, (1988).

8. Hale, A.R., Journal of Occupational Accidents, 6, 17-33 (1984).

9. Hale, A.R., and Hale, M.D., Review of Industrial Accident Literature, National Institute for Industrial Psychology, London (1972).

10. Reason, J., In: Human Factors in High Risk Situations, Royal Society, London, (1989).

11. Withers, J., Major Industrial Hazards, Gower, London, (1988).

Safety education in industry

Dr G.T. Clegg

Department of Chemical Engineering, University of Manchester, Institute of
Science and Technology, Manchester, M60 1QD, United Kingdom

Summary - The presentation reviews some of the major hazards associated
with the storage of flammable liquids, where well-understood, costly
accidents recur with depressing regularly. The value of regular, well-prepared
safety education meetings is discussed against this background.

STORAGE OF FLAMMABLE LIQUIDS

Losses through fires in storage constitute a major part of the losses sustained by the oil and
chemical industry. For the most part, such fires result only in financial loss, but there are exceptions
such as the Feyzin and Mexico City disasters.

Consider some of the hazards involved in the storage of liquefied petroleum gas (LPG) under
pressure. LPG is a generic term for propane (C_3H_8) butane (C_4H_{10}) and mixtures of the two in their
liquid form. Figure 1 gives some of the properties of LPG. In very large quantities, LPG is stored at
near-atmospheric pressure in refrigerated storage tanks, but in quantities of less than 5000 tonnes
it is usually stored under pressure in "bullets" or spheres. In pressure storage installations, the
major source of hazard continues to be the so-called Boiling Liquid Expanding Vapour Explosion
(BLEVE). Figure 2 illustrates the mechanism of a thermally induced BLEVE, and underlines the fact
that adequately sized pressure relief systems will not prevent explosions when exposure of the tank
to fire weakens the shell of the storage vessel (ref. 1). In addition to the large fireball which is
normally generated, the explosion over-pressure can cause shifting of near-by tanks. If pipe failure
results, the accident can then propagate. This seems to have occurred in the Mexico City disaster,
where a large segment of one of the bursting LPG bullets travelled over 1000 m. Even if a tank
doesn't fail, sustained emergency discharge from a relief system can produce dangerous clouds of
flammable gas. Delayed ignition of such a cloud can lead to a powerful vapour cloud explosion,
particularly if the vapour is partially confined by congested pipe work.

Property	Commercial Butane	Commercial Propane
Atmospheric Boiling Point	$-2°C$	$-45°C$
Vapour Pressure at 20°C	2.75 bar	9 bar
Vapour Density Relative to Air at 15.6°C and 1015.9 mbar	1.90–2.10	1.40–1.55
Ratio of Gas to Liquid Volume at 15.6°C and 1015.9 mbar	233	274
Limits of Flammability	1.8–9.0% v/v	2.2–10% v/v

Figure 1
Typical properties of LPG

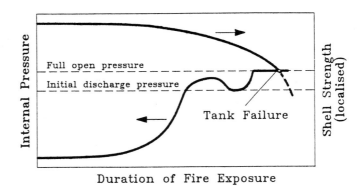

Figure 2
Typical time–pressure–strength relationships
for tanks exposed to fire (From ref. 1)

The safety philosophy for the prevention of BLEVEs usually concentrates on:
a) Minimising the possibility of spillage, which can result from overfilling, water draw-off operations, damage to pipework etc.
b) Avoiding the formation of liquid pools beneath pressure storage tanks by the provision of proper drainage.
c) Reducing the rate of tank temperature increase during a fire by the use of fireproof insulation and water drench systems.

It is surely axiomatic that all concerned with pressure storage of LPG should be aware of the hazards outlined above, as well as the particular prevention philosophy adopted by their company. In addition, operators should clearly understand that:

1) An LPG spill is much more dangerous than a spill of a flammable liquid which has a boiling point above ambient temperature. In the case of propane, evaporation of one volume of liquid will produce some 250 volumes of vapour which is heavier than air. Since the lower explosion limit of propane in air is 2 % by volume, this can in principle result in 250x50=12,500 volumes of explosive mixture.

2) Great care is required if it is ever necessary to discharge LPG to atmosphere, as for example when drawing water from the bottom of a tank. Flashing and self refrigeration of the liquid will occur, which can lead to freezing or jamming of valves in the discharge line. In the case of propane, the discharging liquid will cool to -45°C, at which temperature it can cause severe 'burns' to operating personnel. An operation of this kind was the trigger for the Feyzin disaster which cost 18 lives.

3) In the event of fire in the vicinity of a tank, no attempt should be made to empty the tank, since the liquid contents will help to keep the vessel walls cool. A burning emission from a tank or pipework should not be extinguished except by cutting off the supply of LPG, since a hazardous vapour cloud may form.

Using carefully prepared material, it should be possible to impart this kind of information to all concerned with the operation, and it is strongly recommended that this is done. It has however been found that people learn better by discussing than by listening, and so something other than simple presentation is required if repeated incidents are going to be reduced in number. Kletz (ref. 2) has suggested that a valuable technique is to discuss incidents that have occurred in the past. In this method, an accident is briefly described to a group of 12 - 20 managers, supervisors or operators and illustrated by slides. The group then question the discussion leader to establish the facts, and then go on to say what they think should be done to prevent similar incidents happening again. The group thus plays the part of the investigating pannel, and the discussion leader plays the part of the witnesses. In this way the group participates in a re-enactment of the original investigation, and it is impressed on their memories much more deeply than it would be if they merely read about it or listened to a talk about it.

The author was involved in the investigation of an LPG accident which would provide a suitable example for such a discussion. In this incident a butane road tanker had arrived at a storage depot in order to unload. The normal unloading procedure involved first equalizing the pressure in the vapour spaces of the road tanker and the storage tank by connecting a vapour line between the two. The suction hose of the unloading pump was then connected to a self-sealing coupling fitted to the tanker discharge pipe. The main tanker outlet valve was located within the tank, and was hydraulically operated as indicated in Figure 3. When a small leak was noticed on the self sealing coupling, the driver sent for a maintenance engineer. This man decided to replace the seats in the coupling before the tanker was emptied, and without checking that the main outlet valve was closed. In fact it was open, and when the coupling was dismantled, a full bore discharge of flashing liquid butane (atmospheric boiling point -2°C) poured into the unloading bay. Panic ensued, and it was some time before an attempt was made to close the main valve. This was found to be jammed open, probably as a result of freezing of water absorbed from the atmosphere into the hydraulic fluid. A diesel-driven fire pump was then started to service emergency water sprays located over the loading bay. Ironically, this was the source of ignition of butane vapour which drifted under a badly fitting door into the pump house. In the resulting fire, several company employees were badly burned. An incident of this type provides much material for discussion and is likely to remain in the memory of the participants for a very long time.

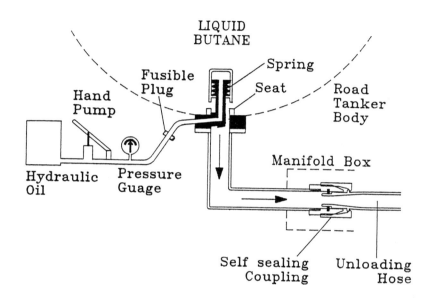

Figure 3
Butane Tanker Unloading System

Many other LPG accidents, including the Feyzin and Mexico City disasters are included in hazard Workshop Modules produced commercially by the British Institution of Chemical Engineers. These modules contain case notes, slides, sometimes videos, and cover subjects including LPG handling, Hazards of plant modifications, Safer piping etc. In a recent paper (ref. 3), a large US company reports that it has periodically presented one of these modules (Hazards of over and underpressuring vessels) to over 700 employees.

REFERENCES

1. W.E. Martinsen, D.W. Johnson and W.F. Terrell, <u>Hydrocarbon Processing</u>, 141 (1986)
2. T.A. Kletz, <u>Education in Chemistry</u>, 7, 229 (1970)
3. R.E. Sanders, D.L. Haines and J.H. Wood, <u>Plant Operations Progress</u>, <u>9</u>, No 1, 61 (1990)

Panel 3: Future Priorities in IUPAC Activities on Safety in Chemical Production

MODERATORS
D. Wyrsch and A. Fischli

There was a full and detailed discussion on this subject with a wide range of views represented. It was agreed that it is an important area for IUPAC and that they could make significant contributions to the area. The importance of 'Responsible Care' was emphasised and the need for IUPAC to be aware of developments in this area. A review of the comments is incorporated in the final summary discussion.

Panel 4: Powder Explosion Protection

MODERATORS
P. Rüedi and R. Siwek

CASE STUDY ON EXPLOSION PROTECTION

There is an inherent dust explosion hazard when combustible dusts are handled in industry. This is proven by many explosions which have resulted in substantial material losses and often in fatalities.

Explosion protection means: (1) Assessment of the probability and effect of explosions resulting from the handling of materials which may generate dangerous and explosible atmospheres. (2) Assessment of the effectiveness of the protective measures which may either be designed to prevent any explosion or to limit the damage in case of an explosion to an acceptable level.

The case study shows how the assessment is applied to determine whether a dust explosion hazard exists in a given manufacturing plant. In the course of the analysis, the following questions have to be raised:

- Will there be an explosible dust atmosphere in the area of the installation to be analysed or within the equipment?

- What is the amount of explosible dust atmosphere present or possible, based on local or manufacturing conditions, and where will it be generated?

- Is the amount of expected explosible dust atmosphere hazardous?

If the questions are answered positively, a decision has to be made with regard to the selection of the protective measures against the imminent dust explosion. A distinction has to be made between two groups:

Group I: Preventive measures which exclude an explosion by:

1) measures which will prevent or limit the formation of explosive atmospheres.

2) inerting or removing oxygen, and

3) eliminating effective ignition sources.

Group II: Design measures which will limit the effects of explosions, they include:

1) explosion-resistant design for the maximum explosion pressure.

2) explosion-resistant design for the reduced maximum explosion pressure in conjunction with explosion pressure venting.

3) explosion-resistant design for the reduced maximum explosion pressure in conjunction with explosion suppression, and

4) means to divert the explosion from entering peripheral equipment.

Explosion protection by **design** measures is always required whenever the goal of avoiding explosions through the application of **preventive** measures cannot be reached with an adequate safety margin. It has to be made certain that no personnel will be injured and that the protected equipment generally will be useable shortly after the explosion.

Diversion devices (group II, # 4) are always necessary in systems which are made up of equipment using preventive explosion protection (group I) and protection by design measures (group II). In the latter, effective ignition sources are anticipated which may result in an explosion that has to be isolated from the other equipment.

With the help of the case study it can be shown, that the presence of explosible dust/air mixtures in a given manufacturing installation does not necessarily represent an explosion hazard although all combustible organic and metallic dusts are capable of exploding. The goal of the case study is to formulate possibilities for the prevention of dust explosions at all or to limit the effects to an acceptable level.

BASIC DATA FOR CASE STUDY "GRINDER"

QUESTIONS:

1) Which data (safety data) of the products to be ground must be knwon?

2) Which additional information is required for planning the technical explosion protection measures?

ANSWERS:

	Information available for continuation of the case study
1) - The following product data must be knwon:	
. Dust explosion class	St 1
. Minimum ignition energy	"low" (<500 J)
. Burning class	2 (20°C) 3 (100°C)
. Decomposition temperature (SPS-test Nr. 5)	160°C
. Shock sensitivity	no detonation
. Content of flammable solvents	none

- The following additional data are available:

. max. explosion pressure P_{max} = 8,0 bar

. K_{St}-value K_{St} = 90 bar.m.s^{-1}

. Minimum ignition energy E_M ≤ 10 mJ

. critical oxygen concentration $(O_2)_{crit.}$ = 10 vol%

2) - The following additional information must be known:

. Is the installation of a pressure schock resistant design?   - no

. Is it a multi purpose or mono plant?   - multi purpose

. In which type of container is the product supplied?   - fibre drum

. In which type of container is the product collected?   - fibre drum

For further planning it is assumed that only St 1-dusts are to be handled in this plant.

IDENTIFICATION OF AREAS WITH EXPLOSION HAZARDS

QUESTIONS:

3) Where have ignition sources to be expected as a matter of principle (positions 1-14 in scheme, p. 180) and what is the nature of each source?

4) When is an explosive atmosphere possible in the different areas?

5) Where could an explosion be initiated and which areas would be affected?

ANSWERS:

Position	Question 3	Question 4	Question 5
1 Charge funnel	electrostatic discharge	during charging	1, 2, vicinity
2 Pre-crusher	-	-	-
3 Intermediate container	-	during charging	-
4 Rotary valve	-	-	-
5 Magnetic separator	-	always	-
6 Hammer grinder	impact and friction sparks, overheating	always	4, 5, 6, 7, 8, 12, 13, 14, ventilation ,vicinity
7 Silo	sparks from grinder	always	4, 5, 6, 7, 8, 12, 13, 14, ventilation, vicinity
8 Rotary valve	-	-	-
9 Flexible hose	-	during discharging	-
10 Fibre drum	electrostatic discharge	during discharging	8,9,10, vicinity
11 Pipeline to filter	-	always	see positions 7, 13
13 Bag filter	electrostatic discharge	always	12,13,14, 8,7,6,5,4
14 Blower	impact and friction sparks	in case of defective filters	14,13,12,8,7,6,5 4,1,Ventilation, vicinity

SELECTION OF THE PROTECTIVE MEASURES

QUESTION:

6) Which protective measures would be adequate for the different areas with explosion hazards or for the entire installation?

ANSWERS:

Protective measure	areas			entire installation
	charging area	silo grinder-filter	discharge area	
Inerting	-	-	-	Charging and discharging devices to be modified, N_2-monitoring in recycling gas stream, decomposition in N_2 atmosphere possible?
Avoidance of ignition sources	X	-	X	-
Pressure shock resistant for the full explosion pressure	-	X Explosion barriers !	- (X)	-
Explosion relief venting	-	X explosion barriers !	(X) (With pressure shock resistant enclosure)	-
Explosion suppression	-	X Explosion barriers! Can the product be suppressed?	-	-
Explosion barriers	X	X Ventilation	X	

For the further planning stages it was agreed to select the protective measures "relief venting" for the areas grinder/silo/filter.

REALIZATION OF THE SELECTED PROTECTIVE MEASURES

QUESTIONS:

7) What are the consequences of the selected protective measures for the protected area and for the remaining parts of the installation?

8) Which information/data must be available for relief venting?

9) Which measures have to be taken to avoid propagation of the explosion into unprotected parts of the plant?

ANSWERS:

7) - The protected parts of the installation must have a minimum pressure (shock) resistance; otherwise they have to be reinforced.

   - Explosion-propagation into unprotected parts must be prevented.

   - For the charge and discharge areas, the protective measure "avoidance of ignition sources" will be applied.

   - The area around the blower (which is an ignition source) is protected by the measure "avoidance of explosive concentrations".

8) - The pressure (shock) resistance of the parts in question - as they are or after improvements - must be known.

   - Length of the relief pipe.

   - Volumes of the protected parts (silo, pipeline to filter, filter).

   - Maximum size of the relief area which could be installed with the existing geometry of the plant.

9) - Replacement of pos. 4, 8, by rotary valves which have been tested and approved (breakthrough of flames, pressure shock resistance).

   - Approved rapid action valve ahead of the blower.

   - Change of air piping so that there is no direct contact with the room where the grinder is installed.

   - Emergency cutout switch to shut down the entire installation in case of activation of the relief device.

   - Safety filter ahead of the blower the exclude an explosive atmosphere even if the main filter es defective.

DESIGN OF THE VENTING SYSTEM

10) Plant data, as determined.

Pos.	Object	Volume	Pressure schock resistance
5	magnetic metal separator	$0,05 \ m^3$	3 bar gauge
6	hammer grinder	$0,02 \ m^3$	>3 bar gauge
7	silo ϕ 2500	$9,0 \ m^3$	after reinforcement 2 bar gge.
12	pipe ϕ 800	$1,4 \ m^3$	~2 bar gauge
13	bag filter, ϕ 1,8 m	$9,2 \ m^3$	after reinforcement~2 bar gge.
14	blower	~$0,02 \ m^3$	~0,5 bar gauge

- Required length of relief pipe: L = 8 m

- The maximum size of the relief area which can be accomodated on the silo is ϕ 1,4 m; but the relief pipe cannot be installed in a right angle to the relief area, but with a deviation of 20° (without bends!).

11) The size of relief area
     (according to the nomograms of the VDI-Guideline 3673, page 179)

- Acceptable reduced explosion pressure: P_{red} = 2 bar

- Volume to be considered   V   = 19,6 m^3

- Influence of the relief pipe: It follows from fig. 9 of VDI-Guideline 3673 that, to obtain a reduced explosion pressure of 2 bar gauge (3 bar absolute) with a relief pipe of 8 m, the relief area has to be sized for a reduced explosion pressure of P_{red} bar gauge (1,31 bar absolute) assuming free relief (i.e. without a relief pipe).

- Static activation pressure: Due to the low value of P_{red}, a static activation pressure P_{Stat} = 0,1 bar gauge (1,1 bar absolute) is selected.

- Required relief area: F = 1,65 m^2

  . Diameter   D = 1450 mm

- Selected: D = 1400 mm

SIZING OF RELIEF AREAS

12.) Original nomograms out of VDI-Guideline 3673 "Relief venting of dust explosions".

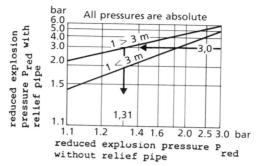

Fig. 9: Influence of the relief pipes on the reduced explosion pressure P_red in the vessel to be protected

<u>Determination of relief areas for a static activation pressure</u>

<u>of 0,1 bar gauge</u>

absolute pressures!

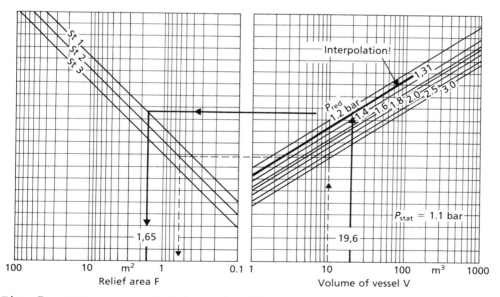

Fig. 7a: Nomograms related to the different dust explosion classes (P_Stat = 1,1 bar)

SCHEME OF THE PROTECTED INSTALLATION

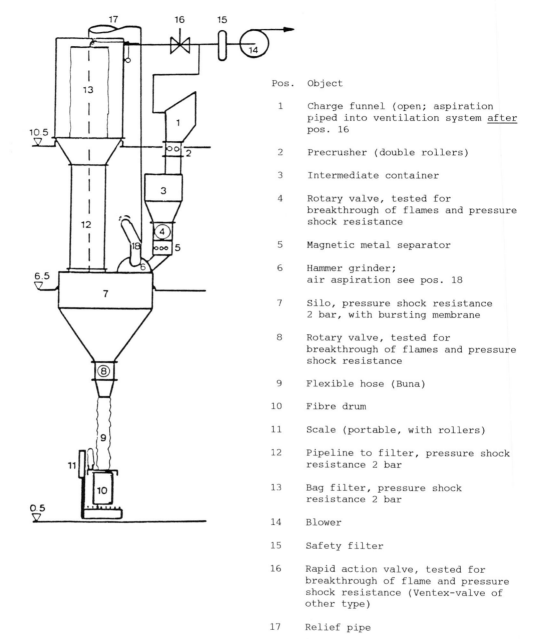

Pos. Object

1   Charge funnel (open; aspiration piped into ventilation system after pos. 16

2   Precrusher (double rollers)

3   Intermediate container

4   Rotary valve, tested for breakthrough of flames and pressure shock resistance

5   Magnetic metal separator

6   Hammer grinder; air aspiration see pos. 18

7   Silo, pressure shock resistance 2 bar, with bursting membrane

8   Rotary valve, tested for breakthrough of flames and pressure shock resistance

9   Flexible hose (Buna)

10 Fibre drum

11 Scale (portable, with rollers)

12 Pipeline to filter, pressure shock resistance 2 bar

13 Bag filter, pressure shock resistance 2 bar

14 Blower

15 Safety filter

16 Rapid action valve, tested for breakthrough of flame and pressure shock resistance (Ventex-valve of other type)

17 Relief pipe

18 Air aspiration of grinder from relief pipe

Additional measures:

- Emergency cutout when bursting membrane blows

- Deluge type water spray nozzles installed in the interior of pos.7.

Panel 5: Static Electricity

MODERATORS
W. Zaengl and M. Glor

CASE STUDIES ON ELECTROSTATICS

The case studies should serve as an aid for the assimilation and deeper understanding of the rules, guidelines and measures which should commonly be applied to prevent ignition hazards due to static electricity in industry. All the cases presented are based on real incidents that occured in production facilities. Some of them caused serious injuries and/or considerable material damage.

The construction of the case studies and the instruction method used are towards the following goal: After working through these case studies the receiver should be able to understand the basic rules and principles for the elimination of hazards caused by static electricity and to apply them in production facilities.

All case studies are set up according to the scheme shown in Figure 1.

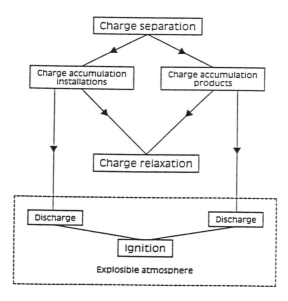

Fig. 1:   Schematic presentation of the different stages in the sequence of events leading to a fire or explosion initiated by discharges of static electricity

First the incident has to be analysed and interpreted:

- describe the separation process
- identify location of charges
- describe discharge path
- identify type of discharge
- estimate ignition energy
- state what additional information is required

Next the preventive measures must be set up:

- equipment
- technology
- organization

The questions marked * have a more theoretical nature. To answer them will contribute to the understanding of electrostatics. But this is not a must. Depending on the time available and on the educational level, treatment of these questions may be omitted.

The questions on preventive measures aim for the essential minimum. Obviously, for every case different levels of improvment are possible, up to the total elimination of the hazardous process or activity.

APPARATUS/EQUIPMENT:

AGITATED REACTION KETTLE

SITUATION:

- Acetone precharged at room temperature
- Stainless steel funnel inserted into manhole

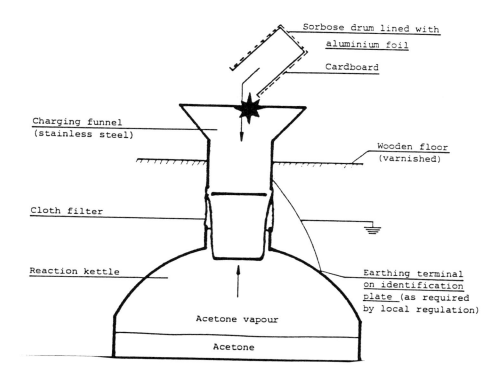

INCIDENT:

When sorbose powder was poured from the drum into the funnel, an explosion occurred in the charge opening.

ADDITIONAL INFORMATION:

· Drum: Capacity of the aluminium foil liner approx. 50 pF.

· Floor: Varnished wood. Earthing resistance 10^7 to 10^8 Ω.

· The operator emptying the sorbose drum was wearing shoes with an earthing resistance of the soles $> 10^8$ Ω.

· Simulation tests after the incident showed that the sorbose drum, when emptied (125 kg of sorbose), carried a charge of approx. 3000 V.

QUESTIONS:

1) What is the separation process that caused the electrostatic phenomenon described?

2) Where is the charge after separation?
   Which parts are charged?

3) Describe the discharge process.
   Where does the charge go?

4) Identify the type of discharge.

*5) What is the minimum ignition energy required to ignite the acetone/air-mixture?

*6) What is the quantity of energy liberated by the discharge?
    Is it sufficient to ignite the mixture?

7) Indicate remedial measures suitable to avoid the incident.

SOLUTIONS:

QUESTIONS:

1) What is the separation process that caused the electrostatic phenomenon described?

2) Where is the charge after separation?
    Which parts are charged?

3) Describe the discharge process.
    Where does the charge go?

4) Identify the type of discharge.

*5) What is the minimum ignition energy required to ignite the acetone/air-mixture?

*6) What is the quantity of energy liberated by the discharge?
    Is it sufficient to ignite the mixture?

7) Indicate remedial measures suitable to avoid the incident.

SOLUTIONS:

1) Separation of sorbose powder from aluminium foil liner.

2) The charge spreads over the aluminium foil (50 pF) and over the insulated operator (150 pF).
    The complementary charge on the sorbose powder drains to earth via the funnel or via the acetone and the kettle.

3) The aluminium foil and the operator are discharged via the earthed metal funnel.

4) Spark

5) 1.25 mJ

6) $W = \frac{1}{2}CU^2 = \frac{1}{2} (50+150).10^{-12}$ F.3000 $V^2 = 10^{-4}$ J = <u>0.9 mJ</u>

    The answers to questions 5 and 6 would indicate that the energy liberated by the discharge would not be sufficient to cause ignition. However, it must be borne in mind that the values given for G and U have a large margin of error. Teh value given for the minimum energy of acetone was taken from literature; it could be lower in an actual case.

7) - Aluminium foil of the drum to be earthed and bonded with the earthed funnel.

    - Floor and shoes must be conductive.

    - Pour sorbose in portions (max. 50 kg) with interruptions of $\frac{1}{2}$ minute between portions (see manual section 3.22 p. 27/28).

APPARATUS/EQUIPMENT

CONAFORM DRIER

SITUATION:

To charge the drier, a charge pipe is inserted into the opening in the floor above the apparatus. The charge pipe had been made of polypropylene to avoid damages to the glass lining of the drier.

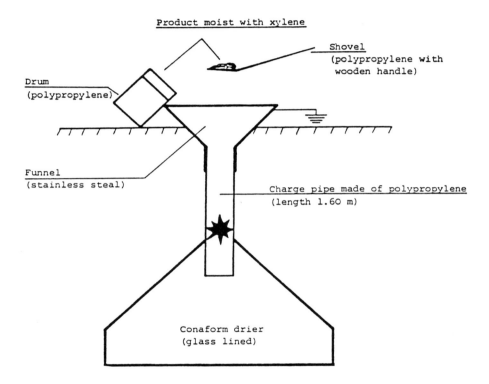

INCIDENT:

When the second drum of a xylene-moist additive for plastics was shovelled into the funnel, an explosion took place in the charge pipe.

ADDITIONAL INFORMATION:

.   The solvent content of the product was 5 to 15%.

.   The flash point of the mixture of xylene isomers is 21 to 30°C.

.   To avoid build up of electrostatic charges, an air humidification device had been installed in the production plant. Nevertheless, relative humidity was only 30 to 50% most of the time.

QUESTIONS:

1) Describe the separation process that caused the electrostatic phenomenon observed.

2) Describe the discharge process. Where does the charge go?

3) Identify the type of discharge.

4) Can the type of discharge (answer to question 4) ignite xylene vapours?

5) Indicate remedial measures suitable to avoid the incident.

SOLUTIONS:

1) Charge separation between xylene-moist product and charge pipe. While charges accumulate on the polypropylene pipe, the charged product falls down into the drier where, presumably, the charges can slowly drain off through the flass lining. However, the charges on the product are not relevant to the incident

2) Charges accumulates on the polypropylene pipe until the charge density per unit area becomes so high that the dielectric strength of the air is reached and discharge takes place even without an earthed conductor being approached.

3) Brush discharge.

4) Yes. Vapours of all flammable solvents can be ignited by brush discharges

5) -Use earthed stainless steel charge pipe.
   -Charge xylene-moist product from conductive drum by means of a conductive and earthed shovel.

APPARATUS/EQUIPMENT:

CYCLONE DUST SEPARATOR

SITUATION:

Due to insufficient conductivity of the gasket bewtween the cone and the flange, the cone is isolated.

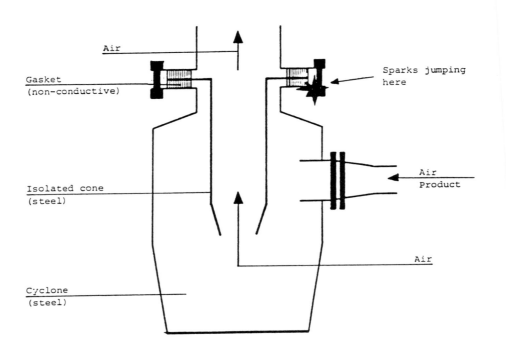

INCIDENT:

When pesticide dust was being separated from air, sparks were seen jumping at the flange holding the inner cone.

ADDITIONAL INFORMATION:

. During overhaul of the plant, non-conductive gaskets had been installed.

QUESTIONS:

1) Describe the separation process that caused the electrostatic phenomenon observed.

2) Where are the charges after separation?
Which parts are charged?

3) Describe the discharge process. Where does the charge go?

4) Identify the type of discharge.

*5) Estimate the capacity of the cone.

*6) What is, normally, the dielectric strength of air?

*7) What is, as a rule, the minimum ignition energy for combustible dusts?

*8) What would be the energy equivalent to a spark with a length of 1 cm in a homogenous field if the capacity is as estimated in the answer to question 5?

9) Indicate remedial measures suitable to avoid the phenomenon observed.

SOLUTIONS:

1) Charge separation between the isolated cone and the dust particles bouncing against it.

2) Charges accumulate on the metal cone (condenser). Complementary charges on the separated dust are drained off via the earthed container.

3) Discharge between the 'collar' of the cone and the nearest earthed part of the equipment (flange, bolt).

4) Spark.

5) 30-50 pF.

6) 3.10^6 V/m (30 kV/cm)

7) 10 mJ.

8) $W = \frac{1}{2}CU^2 = \frac{1}{2}C(Ed)^2 = .40.10^{-12}F(3.10^4 \ .1)^2V^2 = 18$ mJ.

9) Cone to be bonded to the earthed housing of the cyclone either by an earthing cable or by means of conductive gaskets.

APPARATUS/EQUIPMENT:

SUCTION FILTER (NUTCHE)

SITUATION:

The feed line became plugged when a suspension of crystals was charged into the nutche that had previously been inerted by CO_2. The nutche was opened to unplug the line. Then it was closed again and flushed with nitrogen for 2 or 3 minutes before filtration was resumed.

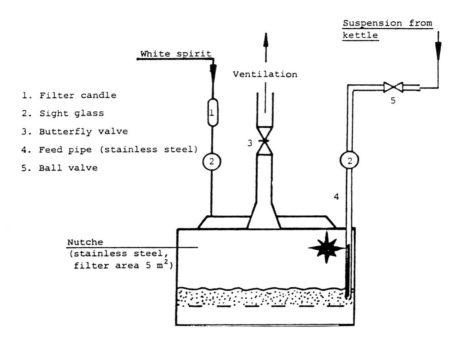

1. Filter candle
2. Sight glass
3. Butterfly valve
4. Feed pipe (stainless steel)
5. Ball valve

Nutche
(stainless steel,
 filter area 5 m^2)

INCIDENT:

When charging of the suspension of crystals (in white spirit) was resumed, a muffled report was heared from the nutche, the cover was thrown open and fire broke out.

ADDITIONAL INFORMATION:

.   The suspension of crystals is very concentrated. It consists of 40% wt of a bacteriostat and 60% wt of white spirit (boiling range 80-110°C, flash point -24°C).

.   Nominal diameter of feed pipe: 40 mm.

.   Time required to charge a nutche of 4 m^3: 10 to 30 minutes.

.   Specific resistance of the bacteriostat suspension: approx. 10^{11} Ωm.

QUESTIONS:

1) Describe the separation process that caused the electrostatic phenomenon observed.

2) Where are the charges after separation?
   Which parts are charged?

3) Describe the discharge process. Where does the charge go?

4) Identify the type of discharge.

5) What are the average flow velocities in the feed pipe if the nutche ist charged in 1) 10 minutes and 2) 30 minutes?

6) What is the maximum velocity that will not produce dangerous charges?

7) Indicate remedial measures suitable to avoid the incident.

SOLUTIONS:

QUESTIONS:

1) Describe the separation process that caused the electrostatic phenomenon observed.

2) Where are the charges after separation?
   Which parts are charged?

3) Describe the discharge process. Where does the charge go?

4) Identify the type of discharge.

5) What are the average flow velocities in the feed pipe if the nutche is charged in 1) 10 minutes and 2) 30 minutes?

6) What is the maximum velocity that will not produce dangerous charges?

7) Indicate remedial measures suitable to avoid the incident.

SOLUTIONS:

1) Charge separation between product suspension and pipeline. Possibly also due to spraying when suspension leaves feed pipe, due to nitrogen pressure applied for charging.

2) The product leaving the charge pipe is charged. When the product is sprayed, a charged cloud will probably be formed. Complementary charges are drained off via the earthed pipe.

3) The very high charge density on the product stream leaving the pipe will lead to a transfer of charges from the product to the steel housing of the nutche.

4) Brush discharge

5) 5.3 m/s for 10 minutes, 1.8 m/s for 30 minutes.

$$V = v \cdot t \cdot Q \rightarrow v = \frac{V}{t\,Q}$$

6) For suspensions in solvents, no safe velocities can be stated.

7) - Use charge pipe of larger cross section to minimize tendency of the line to become plugged and to eliminate the need for nitrogen pressure. (With gravity feeding through an 80 mm pipe a constant velocity of 1-2 m/s was achieved.)

9) - Inerting to be monitored by measuring oxygen concentration.

   - Improvement of conductivity by means of an antistatic additive (by addition of 250 ppm of an additive, the conductivity could be in-creased to $10^{-8}\Omega^{-1}m^{-1}$).

APPARATUS/EQUIPMENT:

REACTION KETTLE

SITUATION:

The kettle contains a solution in methanol at 60°C. For the addition of a powder, the manhole is opened and the suction ventilation is adjusted so that an explosive mixture will not be present near the manhole.

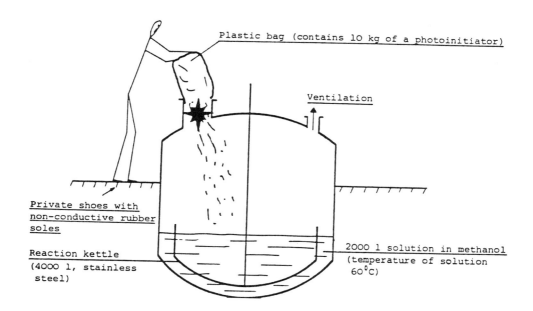

INCIDENT:

A deflagration took place in the manhole area when 10 kg of a powdered photoinitiator were poured from a plastic bag into the kettle.

ADDITIONAL INFORMATION:

- The earthed funnel provided for charging the photoinitiator had not been used.

- During the last 8 years several thousands of batches were prepared without an incident.

- On the day of the incident air humidity was extraordinarily low.

QUESTIONS:

1) Describe the separation process that caused the electrostatic phenomenon. Could there be more than one explanation?

2) Where are the charges after separation?
   Which parts are charged?

3) Describe the discharge process. Where does the charge go?

4) Identify the type of discharge.

5) Could the incident have been avoided with certainty by wearing
   conductive shoes and using earthed metal funnel provided?

6) What was the influence of the air humidity on the incident?

7) Indicate remedial measures suitable to avoid the incident.

SOLUTIONS:

Note: There are two possible causes.

1) -Charge separation between photo initiator and plastic bag
   -Charges accumulating on the operator during walking and manipulating, due to
    non-conductive shoes

2) Charges are on the plastic bag and/or on the operator. Complementary charges on the
   powder are drained off through the methanol

3) -Discharge of the plastic bag at the brim of the manhole
   - Discharge of the operator at the brim of the manhole

4) -Brush discharge
   - Spark

5) No. Brush discharge between plastic bag and an earthed conductor is possible in any case

6) Low air humidity may have been a contributory factor. But increasing the humidity of
   the air would not be a suffiencient measure

7) -Wear conductive shoes and use conductive containers
   -Conductive container and steel charge funnel to be bonded and earthed
   -Product to be poured through the manhole on portions of not more than 50 kg, with
    intervals of at least 30 seconds
   (-Open charging into solvents at temperatures close to their boiling points is not
    permissible.)

APPARATUS/EQUIPMENT:

AGITATED KETTLE WITH BOTTOM VALVE

SITUATION:

The kettle had been cleaned with 25 litres of benzene. When a few litres had been discharged to the drum, the bottom valve plugged. For unplugging the kettle was pressurized with nitrogen to 5 bar.

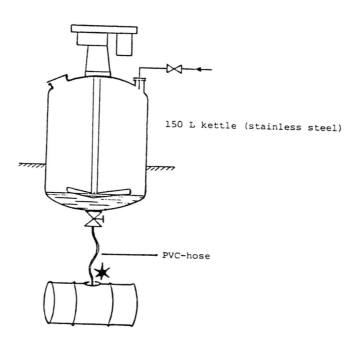

150 L kettle (stainless steel)

PVC-hose

INCIDENT:

During the attempt to unplug the valve by repeated opening and closing motions, the benzene suddenly began to flow and a fire broke out at the bunghole of the drum.

ADDITIONAL INFORMATION:

- PVC has a volume resistance of aporox. 10^{13} ohm.m.

- It can be assumed that the kettle was earthed via the supporting structure.

- It is not known whether the drum was earthed.

QUESTIONS:

1) Describe the separation process that caused the electrostatic phenomenon observed?

2) Where are the charges after separation?
   Which parts are charged?

3) Describe the discharge process. Where does the charge go?

4) Identify the type of discharge?

5) Is the available information sufficient to explain the incident?

6) Indicate measures suitable to avoid the incident.

SOLUTIONS:

1) Charge separation between benzene and PVC hose

2) Charge is accumulated on the PVC hose, opposite charges are in the benzene

3) Discharge between the outer surface of the PVC hose and the brim of the bunghole of the drum.

4) Brush discharge

5) Yes. All conditions for a brush discharge at the bunghole are met. Benzene can be ignited by brush discharges

6) Use a conductive hose (maximum resistance between couplings 10^6 ohm). Ensure conductive coupling of the conductive hose to the kettle. End of hose to be bonded to the drum. (It is recommended to use a vented filling cone for this operation. This device automatically provides good bonding and at the same time improves the hygiene situation since vapours are sucked away). The drum receiving the benzene must be earthed.

APPARATUS/EQUIPMENT:

DRUM FILLING STATION

SITUATION:

The filling nozzle is located 1 cm above the bunghole of the drum. Solvent was to be filled into the drum from an elevated tank.

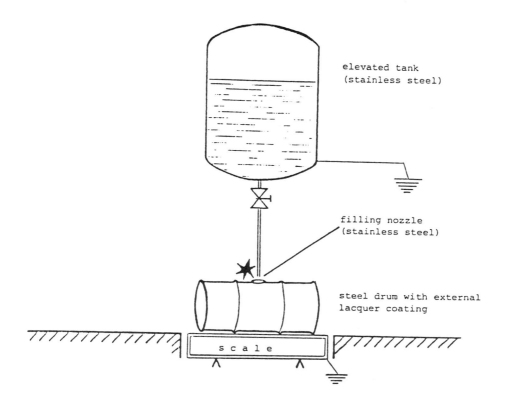

elevated tank
(stainless steel)

filling nozzle
(stainless steel)

steel drum with external
lacquer coating

scale

INCIDENT:

When the second drum was to be filled, a deflagration occurred immediately after the valve had been opened.

ADDITIONAL INFORMATION:

· The tank and the scale were earthed via copper conductors.

· The drums were of the non returnable type with external lacquer coating.

· The flash point of the solvent was below 21°C.

QUESTIONS:

1) Describe the separation process that caused the electrostatic phenomenon observed.

2) Where are the charges after separation?
   Which parts are charged?

3) Describe the discharge process. Where does the charge go?

4) Identify the type of discharge.

5) Indicate remedial measures suitable to avoid the incident.

6) How could you explain that the incident occured with the second drum and not already with the first one?

SOLUTIONS:

1) Charge separation between solvent and filling nozzle.

2) The charge in the solvent leaving the noozle is carried into the drum and accumulates on the drum. Due to the external lacquer coating of the drum, this charge cannot drain away via the earthed scale.

3) Discharge from the isolated drum onto the earthed filling nozzle (distance 1 cm).

4) Spark

5) Drum to be earthed and bonded to the filling nozzle. (Non-conductive coating has to be scraped off prior to fitting the earthing clamp. For durm filling, vented filling cones have been developed. They provide good bonding between filling nozzle and drum and at the same time improve hygiene conditions since vapours are sucked away.)

6) Possibly on the first drum the lacquer coating had worn off at the point where the drum rested on the scale so that the charge could drain away.

APPARATUS/EQUIPMENT:

RAIL TANKER

SITUATION:

Plastic powder is pneumatically conveyed into rail tanker.

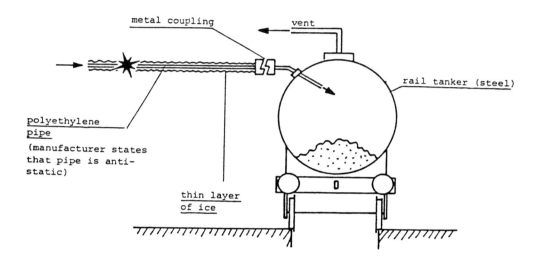

INCIDENT:

Shortly after the onset of pneumatic conveyance a deflagration occurred within the polyethylene pipe. After earthing of the metal coupling (which was thought to have been the cause) the same incident occurs a second time.

ADDITIONAL INFORMATION:

.   Dimensions of the rail tanker:

Length: 13   m   weight: 44 t
     Ø: 2.7 m   volume: 75 m^3

.   The incident occurred on a winter day.

.   After the incident, the surface resistance on the inside of the polyethylene pipe was measured and found to be 10^{12} Ω.

QUESTIONS:

1) Describe the separation process which caused the electrostatic phenomenon observed.

2) Where are the charges after separation?
Which parts are charged?

3) Describe the discharge process. Where does the charge go?

4) Which type(s) of discharge is (are) possible?
Where would a discharge probably occur?

5) Would the type(s) of possible discharge release sufficient energy to ignite a dust/air mixture?

6) Could the incident have been prevented by earthing all conductive parts?

SOLUTIONS:

1) Separation between particles of the plastic powder and the PE pipe.

2) Plastic powder and PE pipe carry charges of opposite polarity. The metal coupling is is also charged, either by the separation process identified above or by influence.

3+4) A spark may jump from the charged coupling to the earthed mass of the rail tanker; depending on geometrical details this could happen either outside or in the interior of the pipe. If there is a layer of ice (conductive) on the PE pipe, a 'Lichtenberg' discharge can occur inside the pipe.

5) Both normal sparks as well as 'Lichtenberg' discharge can ignite dust/air mixtures.

6) No. The pipe should be made of conductive material.

Panel 6: Thermal Analysis

MODERATORS
G. Killé and F. Stoessel

PROCESS THERMAL SAFETY CASE STUDY

INTRODUCTION

The present case study has been conceived by considering actual data. It is devised in such a way that it is possible to get a deeper insight in one or another aspect of the thermal safety of the process.

TOPIC

It deals with the technical and safety study of a stage of a synthesis process. The desired reaction is highly exothermic and the reaction mass is stable only within a very limited temperature range. One intends to build up appropriate reaction conditions based on the stability of the reaction mass and the cooling capacity of the reactor.

The problem consists of three parts that can be treated separately if necessary.

1) THERMAL STABILITY OF THE REACTION MASS

The provided thermal data are used to determine the range of parameter values that defines a safe operating of the reaction process.

2) HEAT TRANSFER

The heat transfert coefficient measures the heat that can be taken off from the reactor. Laboratory measurements and a simple measurement on the production reactor make it possible to calculate the cooling power of the industrial reactor under different possible working conditions.

3) REACTION OPTIMIZATION

The working operation of the reaction (batch/semi-batch) is defined first. It will then be possible to analyze the results of a design of experiments in such a way that the optimal working can be derived by taking into account the different limits.

1. THERMAL STABILITY OF THE REACTION MASS (OVERVIEW)

1.1. DATA

The reaction runs like

$A + B \ ----> \ P \ ----> \ S$

A,B reactants
P desired end product
S decomposition product

Thermal stability of the reactant A

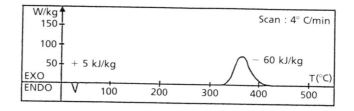

Fig. 1 : DSC Thermogram of the starting product A

The reactant B shows no exotherm below 500°C

Thermal stability of the final reaction mass :

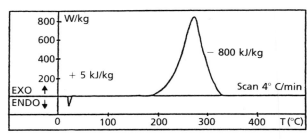

Fig, 2 : DSC Thermogram of the final reaction mass

Both reactants A and B were mixed at the ambient in a pressure resistent steel cell which was then heated linearly in the DSC apparature.

1.2 QUESTIONS

On the basis of this study, do you conclude that this reaction may be performed as a batch reaction (adiabatic or isothermic conditions) ?

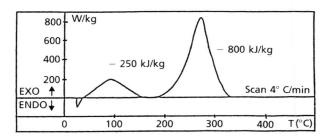

Fig.3 : DSC Thermogram of the cold mixed reactants

Under which conditions could the responsability for such a reaction operating be accepted ?

Which other type of reaction operating may also be selected ?

Which additionnal experiments are then necessary ?

The heat capacity of the reaction mass is 1.7 kJ/kg.K

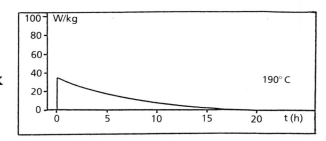

Fig.4 : Isothermic Thermogram at 190°C

2) HEAT REMOVAL

2.1. DATA

It is intended to perform this reaction in a 16m³ semi-batch vessel with external semitoric welded pipes. The chosen inital amount is such that the volume is 15m³ before dosing and the heat exchange area is 20 m². 3m³ of the reactanst B are added.

2.2. QUESTIONS

Calculate the cooling power of the vessel in kW/m³ solution of A at 80,90,100,110 and 120°C. Since the reactant A becomes stiff at 25°C, the minimum temperature of the cooling medium should be 30°C.

One assumes a value of 300 W/m².K for the heat transfer coefficient. The calculation of the cooling power will be advantageously refered to the initial volume of 15m³.

3) REACTION OPERATION

3.1 AIM

With the results of the two first problems (which are recalled below), the optimal reaction conditions are to be worked out with regard to the various border conditions.

3.2. DATA

- Chemistry : The reaction corresponds to the following scheme
  A + B ----> P
  A is in solution (2kmol/m^3) and is introduced as it is.
  B is in 11 molar solution. We know from probl.1 that P decomposes into S,
  and P ---> S (further reaction)

- Stoechiometry : On an ecological and economical basis, the excess of B is set equal to
  10% ; hence, the molar ratio B/A = 1.1

- Temperature : In problem study 1, the maximum temperature (that is allowed to be reached
  in case of trouble) was set at 130°C.

- Reactor : A stirred reactor of 16m^3 nominal volume will be used. The maximum
  possible volume is 20 m^3 and the 20 m^3 maximum cooling area corresponds
  to a volume of 14 m^3.

- Quantities : The present synthesis stage is one of several ; therefore, the quantities must
  be adjusted to those of the preceeding and following stages. Thus, the
  resulting volume of solution of A is 15 m^3 and the corresponding volume of
  solution of B is 3 m^3.

- Cycle Time : In order to synchronize this synthesis stage with the other stages of the
  synthesis sequence, a cycle time of 12 hours must be held. Therefore, by
  taking into account the heating up and transformation times, 10 hours are left
  for the reaction time its elf.

- Quality : The conversion should be 99%

- Cooling : (from Problem 2)

Reaction operating

The solution of Problem 1 rejected the batch reaction operating. Because the continious reaction operating was not considered to be possible for technical reasons, the semi-batch reaction is left. Owing to the selectvity of the reaction and the ratio of the volumes of the two reactants, A has to be introduced first in the vessel and B is added to A.

3.3. QUESTIONS

Suggest a design of experiments that will allow you to work out an operating procedure. One should try especially to define the adaptable parameters with the corresponding variation limits and border conditions

SOLUTION OF THE THERMAL SAFETY PROBLEM

1) THERMAL STABILITY OF THE REACTANT A

The decomposition of A occurs within a temperature range ($\approx$300°C) which cannot be reached by itself .Would the main reaction lend to such a temperature, there would be no more A left but the final reaction mass would be present.

CONSEQUENCES

Effect of an event :low   (-60 kJ/kg)
                    Tad 35°C

Probability : low, because decomposition occurs at a higher temperature only

FINAL REACTION MASS

The decomposition energy is huge : 800 kJ/kg; this corresponds to an adiabatic temperature raise of about 470°C. The temperature at which this decomposition reaction is observable in the dynamic DTA experiment is low, about 160°C.

The question now is : can this decomposition be started when the main reaction runs out of control ? This depends strongly on the way the reaction is operated

BATCH REACTION OPERATING

In this kind of reaction operating, all the reactants are introduced in the vessel and the mixture is heated to the reaction temperature.
The reactor is emptied when the reaction is over. With such a procedure, among others two particular temperature operatings are possible : adiabatic and isothermal.

o ADIABATIC REACTION OPERATING

The reaction energy is -250kJ/kg, hence, for a starting temperature of 30°C, a temperature of 30x250 /1.7 = 177°C, ≈180°C can be reached.

The question is now : is the decomposition reaction already dangerous at 180°C ?

Fig.4 shows that the heat evolved of the reaction is 40 W/kg.By using Van't Hoff's law one gets

40 W/kg at 190°C
2O W/kg at 180°C

If one assumes that the final reaction mass is at 180°C (final temperature by adiabatic reaction operating started at 30°C) and that the cooling cames to break down how will this reaction mass behave ?

Under adiabatic conditions, the reaction heat is entirely used to raise the temperature. The rate of the temperature raise is :

$$\frac{20\text{W/kg}}{1700\text{J/kgK}} = 11.8.10^{-3}\text{K/s or 42 °C/h}$$

To go from 180 to 190°C, it takes 0.24 h with 20 W/kg
To go from 190 to 200°C, it takes 0.12h with 40W/kg
and so on..

The runaway takes place within ≈ 1/2 h

In this calculation, it was nevertheless assumed that the reaction energy flow remains equal to 20 W/kg from 180 to 190°C , and hence, the time has been overestimated.

CONSEQUENCE

These times are very short from an industrial point of view, one may thus strongly advise against such on adiabatic procedure.

o ISOTHERMAL REACTIVE OPERATING

For safety needs, we consider the highest occurable accident. In this case, there is a cooling break down at the most critical time, i.e. at the beginning of the reaction. This brings back to the adiabatic reaction process.

DETERMINATIONS OF THE HIGHEST ALLOWABLE TEMPERATURE

A series of isothermal DTA measurements at different temperatures allow to determine the maximum heat flow as a function of temperature. The analysis of the flow by using Arrhenius law (see below) leads to the activation energy of the decomposition.

From these data, the so called TMR-Formula provides the induction time of the runaway reaction (under adiabatic conditions) and thus the probability of such an event can be estimated.

$$TMR_{a}(s) = \frac{Cp.R.To^{2}}{q_0E}$$

Cp = Heat capacity   J/kg·K
R = gas constant   8,314.J/mol·K
To = initial temperature K
qo = flow at To   W/kg
E = activation energy   J/mol

In our case, the activation energy is 100 kJ/mol

The heat flow of the decomposition and the corresponding TMR ad can be calculated for various temperatures.

Température °C	100	110	120	130	140	150	160
Flow W/kg	0.16	0.36	0.81	1.7	3.6	7.1	13.7
TMR_{ad} h	77	35	17	8.1	4.2	2.2	1.3

The safe working field is defined in such a way that the induction time for the decomposition is at least 8 h. This value revealed to be satisfactory for a "normal superviewed" operation. The relatively large safety margin takes into account the inaccuracy of the extrapolation on the logarithmic diagram.

Thus, in our case, the temperature to be reached must not be higher than 130°C

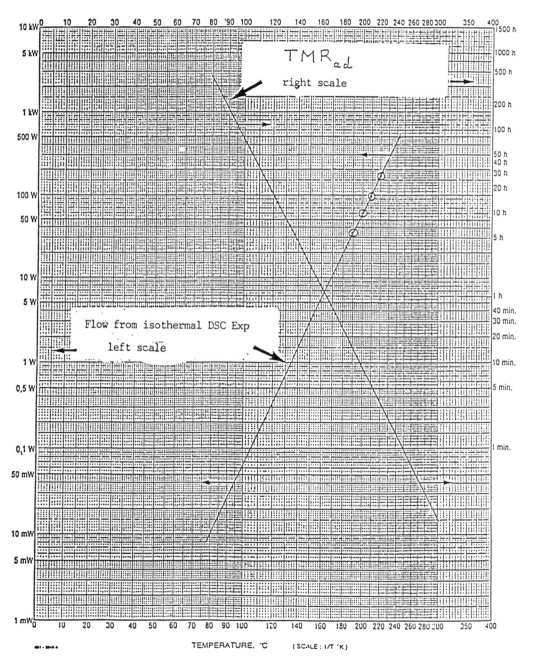

TEMPERATURE. °C   (SCALE : 1/T °K)

FIG. 5

2 HEAT REMOVAL

The previously considered vessel has a maximum exchanging area of 20 m^2 . This value is reached from a volume of 14 m^3. Thus, the heat transfer area remains constant during the dosing. A temperature of 30°C for the heat transfert fluid and a heat transfer coefficient of 300 W/m^2K give the following flows :

Temperature (°C)	80	90	100	110	120
Flow kW/m^2	20	24	28	32	36

3. REACTION OPTIMIZATION

ADJUSTABLE PARAMETERS

After having considered all the border conditions, two more parameters can be adjuste.

. the temperature, from 80 to 120°C
. the dosing time of B, between 2 h (1,5m^3/h) and 8 h.

BORDER CONDITIONS

Conversion rate : 99% in maximum 10 h
Maximum temperature in case of cooling break down : 130°C
The maximum heat flow must correspond to the cooling capacity.

DESIGN OF EXPERIMENTS

We consider here only an example :

5 temperature levels are selected (°C): 80   90   100   110   120
4 dosing times   (h)   :   2   4   6   8

The experiments were performed in a reaction calorimeter.

The results are displayed in three tables

TABLE 1 Reaction time for 99% conversion (h)

Temperature	°C	80	90	100	110	120
Dosing time	2h	>10	7.4	5.0	3.8	3.0
	4h	>10	9.0	8.6	5.4	4.8
	6h	>10	>10	8.4	7.2	6.6
	8h	>10	>10	>10	9.0	8.4

TABLE 2 Maximum heat flow of the reaction mixture (W/1g)

Temperature	°C	80	90	100	110	120
Dosing time	2h	61	61	78	83	86
	4h	36	40	42	44	44
	6h	26	28	29	29	30
	8h	20	21	22	22	23

TABLE 3 Maximum heat accumulation (% of the total heat of reaction)

Temperature	°C	80	90	100	110	120
Dosing time	2h	38	29	22	17	14
	4h	27	21	16	12	9
	6h	22	17	19	10	8
	8h	19	14	11	8	7

The highest temperature that can be reached in case of a control loss of the main reaction can be calculated from the maximum heat accumulation. Thus, TABLE 3 can be transformed into a table giving the maximum final temperatures.

Data :   Reaction heat : $\Delta H_R = 255$ kJ/kg at the stoechiometric point

Heat capacity of the reaction mass $Cp = 1700$ J/kg.K

$$\Delta Tad = \quad 150 \quad °C$$
$$\Delta H_R = \quad 255 \quad kJ/kg$$
$$Cp = \quad 1.7 \quad kJ/kg.K$$

The accumulation ratio gives the maximum reachable final temperature

$$T_{end} = T_R + \frac{\Delta H_R}{Cp}.X_{AC}$$

(X_{AC} = fraction of the accumulatable heat from TABLE 3)

Thus, we build up TABLE 4

TABLE 4 : Maximum reachable final temperature in case of cooling break down.

Temperature	°C	80	90	100	110	120
Dosing time	2h	137	133	133	136	140
	4h	121	121	122	128	134
	6h	113	115	119	125	131
	8h	109	112	116	123	130

After examination of the procedures taking into consideration the three criteria :

- heat flow of the reaction/cooling power
- 99% conversion in less than 10h
- maximum temperature in case of cooling break down : 130°C

The following proceeding variants are still left :

reaction temperature	110 °C	110°C	120°C
dosing time	6h	8h	8h

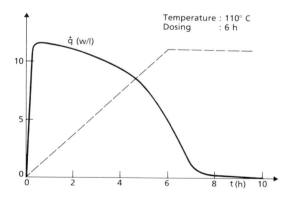 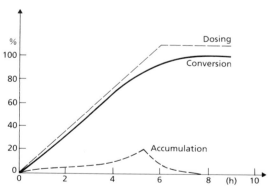

FIG . 6

Panel 7: Occupational Hygiene

MODERATORS
M. Guillemin and R. Roth

Introduction

Hazards in the occupational environment are of multiple nature. Their origins, sources, impacts and consequences on life, health or material may be extremely different and sometimes very complex to assess or to anticipate.

The control and management of these hazards require such a large amount of knowledge and experience in various fields that a multidisciplinary team approach is necessary.

This paper will present the nature and the role of the science called 'occupational hygiene' in the general framework of risk management in the chemical industry. It should be stressed that such a presentation cannot be comprehensive and will be limited to a few relevant key elements aiming at the illustration of the improtance and the usefulness of occupational hygiene in the management and control of hazards. Focus on chemical hazards appears to be appropriate in this context but does not mean that occupational hygiene is restricted to these ones.

The purpose of this paper is to provide the 'educators' in the chemical industry or in universities with a simple document introducing occupational hygiene as a necessary tool for managing health and safety in the chemical production.

Definition of occupational hygiene

The International Association of Occupational Hygiene (IOHA) has adopted the following definition:

'Occupational Hygiene is the discipline of anticipating, recognizing, evaluating and controlling health hazards in the working environment with the objectives of protecting worker health and well-being and safeguarding the community at large.'

This definition helps to understand which type of hazards may be controlled and managed by this science and by which approach. In principle, these hazards and stresses should be measurable so their health risk (both acute and chronic) is concerned. The basic approach described in this definition is divided in three major steps:

1) Anticipation/recognition (detection)
2) Evaluation (assessment)
3) Control (correction or prevention)

Moreover this definition stresses that occupational hygiene is also concerned by the community as a whole. This means that the control of pollutants and hazards inside the factory must take in to account the protection of the environment by eliminating (substitution, retention, destruction) the potential air, water or soil contaminants at their source instead of emitting them in the environment. An appropriate risk analysis and hazard evaluation at the workplaces will also help to avoid catastrophic events and therefore will protect the surrounding population. This important aspect of occupational hygiene has often been underestimated.

'Occupational hygiene is a science which protects both the workers and the environment.'

According to a document of the World Health Organization [1], 'an occupational hygienist is a person with a university degree in engineering, physics or chemistry or an equivalent science degree (in some contries, medicine), and in addition with specialized training in recognition (identification of hazards and understanding of their effects on health of humans and their wellbeing), evaluation (characterization of hazards from qualitative and quantitative points of view) and control (preventive measures) of hazards that arise in or out of the workplace and may cause impaired health or significant discomfort among workers or inhabitants of the surrounding community'.

Role and function of occupational hygiene

Safety usually refers to the accident prevention and hygiene to the prevention of diseases or discomfort. This difference between safety and hygiene, although useful in designing different fields of knowledge, may also appear somewhat abritrary. Safety may not be limited to fire, explosion or accident prevention and may encompass also prevention of long term hazards leading to chronic diseases. On the other side, occupational hygiene may include in its activities fire, explosions or accidents prevention and certainly includes the control of acute risk such as intoxication, narcosis or strong irritation which are accidents. Occupational medicine has the same objectives as occupational hygiene (protection of the worker's health) and uses a complementary approach by focusing its attention at the worker in order to ensure that his/her job fits his/her specific physiological and psychological conditions and will not impair his/her health with time.

From the above it is clear that occupational hygiene holds quite a privileged position in the multidisciplinary team in charge of controlling and managing the hazards in the chemical industry. The occupational hygienist who focus his/her attention at the workers' environment finds in the occupational physician the necessary partner to reach their common goal. He/she is also a privileged partner of the other persons in charge of safety or environmental protection since the problems encountered in the occupational environment often concern simultaneously hygiene, safety and ecology. In other words, the occupational hygienist play and important role of coordination between the different disciplines of such teams and may also act as a catalyst to the cross-fertilization which occurs in a multidisciplinary approach of the problems.

Figure 1 illustrates one of the aspects of the privileged position of occupational hygiene in the field of workers' health protection [2].

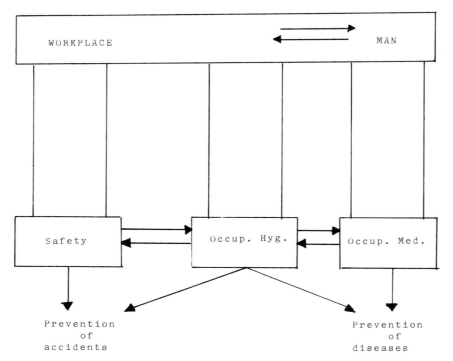

Figure 1 Framework of the workers' health protection

Scope and function of occupational hygiene is obviously not limited to the three steps described above, but include a lot of other important tasks such as advice on planning and organization of work, participation in the developments of programme to improve the workplace conditions, training and education of all the people involved, dissemination of information, etc....[3]

The role of occupational hygiene in the process of risk assessment and in the field of risk management is determining and essential in the chemical industry because the majority of the hazards are of chemical nature and concern the problems specific to that science which is obviously the most appropiate to keep these risks under control.

Basic approach of a specific problem

By the way of a typical situation in a chemical industry we will illustrate how an occupational hygienist will approach a given problem. It should be stressed again that this example will only present a few very limited aspects of the role of an occupational hygienist.

The situation to investigate is the production of chemical , in a typical production line consisting of different vessels (mixture, reaction,...) and units (filtration, drying,...).

The problem to solve is to know if the health risk due to chemicals for the workers occupied with this production in its normal use (abnormal events or accidents being another problem here) is acceptable or not.

(a) Anticipation-recognition

The hygienist must gain accurate knowledge about this manufacturing process and collect information about all the chemcials involved in this production including by-products, impurities, wastes, etc.

Beside the usual physical and chemical properties of the substances such as flash point, explosive limits, flammability, evaporation rate, vapour pressure, and so on, other useful informations should also be collected (Table 1), as well as an estimation of the amounts involved.

Through careful observations and surveys of the workplaces, through qualitative preliminary s hazard analysis and with the help of all the other relevant information available (inspection or audit reports, incident or accident statistics and analysis, routine monitoring of the process, medical information about the workers, etc....), the hygienist and his/her team composed of the concerned persons (production manager, occupational physician, chemist, process engineer, foreman, etc....) will identify the workplaces where potential health risks do exist.

This first step may be quite difficult to achieve, but is of utmost importance since any undetected hazard will remain unnoticed until a deleterious health effect will be observed with the workers (fail of the prevention).

Table 1 Useful informations related to chemical substances for assessing their health risk

----

Acute	Chronic
Minimum anesthetic concentration	Solubility in water and fat (oil)
Irritation treshold	Metabolism pathways/clearance
Chemical reactivity	routes
Odor threshold	Pharmacokinetic/pharmacodynamic data
Level immediately dangerous to health	Permissible exposure level
Short term exposure limit	Biological exposure indices
Lethal dosis	Acceptable daily intake
.....	....

Note: some information may be related to both acute and chronic risks.

(b) Evaluation-risk assessment

In our example, the risk assessment will rely on the exposure assessment since we have restricted the problem to the chemical hazards.

The hygienist will use his/her professional judgement for each situation with a potential health risk, to dertermine if the risk is obviously so low that it can be accepted without further investigations or if it is obviously so high that immediate corrective action is needed. For all the other situations where a clear decision cannot be taken, the risk (exposure) has to be evaluated. Three methods come into consideration for this evaluation [4].

   (i) Ambient monitoring is the measurement and assessment of agents at the workplace evaluate ambient exposure and health risk compared to an appropriate reference.

   (ii) Biological monitoring is the measurement and assessment of workplace agents or their metabolites either in tissues, secreta, excreta, expired air or in any combination of these to evaluate exposure and health risk compared to an appropriate reference.

   (iii) Health surveillance is the periodic medico-physiological examinations of exposed workers with the objective of of protecting health and preventing occupationally related disease.

These three methods have their own advantages and limitations and should be used simultaneously whenever possible. However, it should be emphasized that the biological monitoring is, at the moment, limited to a few agents (approximately less than 40). The health surveillance carried out by the occupational physician is a very important part of the risk assessment because it allows to detect early reversible signs of health effects or biological response (when the appropriate tests and medical investigations have been selected) which may occur in sensitive individuals or in situations where a complex exposure pattern has produced unexpected synergistic effects.

The hygienist's skill for ambient exposure assessment will be used here to set up an appropriate sampling strategy including the selection of an adequate analytical method or the use of a reliable direct reading instrument. The choice of the sampling method (either stationary or personal), the duration of the sampling, the number of workers involved, the locations of the fixed stations, the type of data handling, the possible external and internal interactions, the pharmacological and toxicological properties of the considered chemicals, the relevance of the chosen index to monitor (in case of complex mixtures), the appropriate references, etc.. will be taken into consideration in this very critical phase of the risk assessment [5]. The possibility of skin resorption and of ingestion through contaminated skin (fingers, lips) has to be taken into consideration in order to assess the internal dose received by the worker. This is where the biological monitoring may be quite helpful.

Figure 2 summarizes the interrelationship between the three monitoring methods and their usual associated 'reference values'.

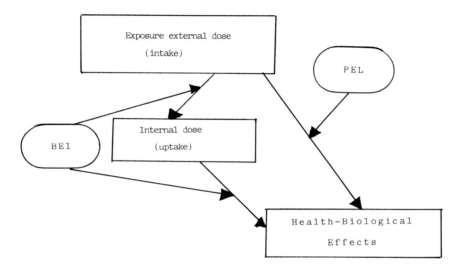

Figure 2 Interrelationship between the different monitoring methods. Permissible exposure level (PEL) and Biological Exposure Index (BEI) are usual 'reference values'.

When the exposure has been reliably assessed the health risk has to be evaluated to know if it can be considered as acceptable or not. When an 'appropriate' reference value does exist, such as a permissible exposure level or a biological exposure index (which may be a legal requirement or a simple recommendation), the interpretation is rather easy since this value represents the acceptable risk. But even in these 'simple cases' the collaboration with a physician is necessary to take in to account the numerous factors which can play a role in assessing the health risk for a given individual (physiological characteristics, life style, drugs consumption, health status and so on).

In the other cases, where the reference value is not straight-forward and estimation may be done by an extrapolation of the other reference values or of the toxicological data such as 'Acceptable Daily Intake' (ADI), 'No Observable Exposure Level' (NOEL), 'Lethal Concentration' (LC), 'Immediately Dangerous to Life or Health' (IDLH) levels, etc. To derive an acceptable risk from these values is a difficult task and requires good knowledge in toxicology and should also be done as a team work after careful considerations and keeping in mind that such assessment will derive much more 'fragile' reference values than those based on good epidemiologic studies.

Our knowledge in this field being yet very poor, a reliable risk assessment becomes rapidly impossible in case of complex mixtures where no information is available about the health effect of this precise mixture or about the possible external and internal interactions (synergism, antagonism...). Prevention, in these cases, implies to minimize the exposure as much as practically possible.

(c) Control and risk-management

If the risk has been found unacceptable, it must be either eliminated (change of substances or process) or diminished to an acceptable level. This 'sanitation' phase is again a team work where the occupational hygienist plays quite an important role. The possibilities to limit the exposure to the workers are usually numerous and should be chosen according to their best best cost/benefit ratio. Automation, enclosure, ventilation, are among the usual preventive measures. The hygienist will test their efficiency and set up a maintenance programme to ensure an adequate protection of the workers.

In some countries, such as in the USA, the occupational hygienists recommend to consider half the permissble exposure value as the 'action level'. In other words above this level, sanitation has to be done. This strategy is based on two important facts:

  (i) Exposure assessment is not always perfectly reliable and higher
       levels are possible.

  (ii) The permissble exposure levels are usually not severe enough since
       most of the time they are corrected on the lower side (lowering of
       PEL).

The risk management [6] involves much more than the risk assessment and the preventive measures described here, but is out of the scope of this paper.

Conclusion

Through a limited and superficial presentation of a few key elements of occupational hygiene, the invaluable contribution of this discipline for the risk management in the chemical industry has been shown.

References

[1] 8th Report of the Joint ILO/WHO Committee on Occupational Health
    'Education and Training in Occupational Health, Safety and Ergonomics'
    Techn. Rep. Series 663, WHO, Geneva 1981

[2] M. P. Guillemin 'Why is industrial hygiene lagging behind?' in Education and Training Policies in Occupational Safety and Health and Ergonomics. Occupational Safety and Health Series No. 47, International Labour Office (ILO), Geneva 1982, p. 59-63

[3] M. Lippman 'Conference working group reports: need scope and function' in Training and Education in Occupational Hygiene: An International Perspective. M. Corn and J. K. Corn (eds). Annals of the American Conference of the Governmental Industrial Hygienists (ACGIH) Cincinnati, Ohio 1988, p. 79-81

[4] A. Berlin, R. E. Yodaiken and B. A. Herman (eds) 'Assessment Toxic Agents at the Workplace'. Commision of the European Communities Martinus Nijhoff Publ., Brussels-Luxemburg 1984

[5] N. A. Leidel, K. A. Bush and J. R. Lynch 'Occupational Exposure Sampling Strategy Manual' US Dpt. of Health, Education and Welfare, National Institute of Occupational Safety and Health, Cincinnati, Ohio 1977

[6] J. T. Garret, L. J. Cralley and L. V. Crolley (eds) 'Industrial Hygiene Management' John Wiley and Sons, New York 1988

Panel 8: Hazards Analysis

MODERATORS
P. Bützer and F. Schmalz

TOPIC OF THE CASE STUDE: ESTIMATION OF CONSEQUENCES

This case study is intended to give an introduction in the way of thinking in scenarios:

- to select a worst-case (or a realistic) accident scenario,

- to determine the different steps following an accidental release of a dangerous liquid,

- to get an impression of how to develop simple models describing these different steps,

- to make a first order estimate on possible dangerous consequences.

The "CASE":

Assume: Among the activities within your responsibility are some operations which necessitate the handling of a dangerous substance. In order to develop an off-site emergency plan, your local authorities want to know the possible consequences of a worst-case accident including this substance. You are asked for a detailed description of conceivable accidents.

Another possibility could be that your insurance company asks for a statement regarding the consequences of the worst-case accident, in order to fix new premiums.

QUESTION:

1.)   What are the most important properties of the substance you
       should know in order to solve the problem?

ANSWER:

1.)   Not only name and chemical formula, but also:

       . physical properties:

              .. molecular weight
              .. state of aggregation
              .. boiling point
              .. vapor pressure
              .. solubility in water
              .. density
              .. volatility
              .. (heat of evaporation)

       . chemical properties:

              .. acidity
              .. auto ignitability
              .. reaction with water
              .. dangerous reactions with other chemicals

       . toxicity:

              .. acute (IDLH-value)
              .. TCLo
              .. LCLo
              .. odor threshold
              .. chronic:
              .. TLW
              .. STEL
              .. carcinogenic

       . ecotoxicity

       . fire and explosion hazards:

              .. flash point
              .. ignition point
              .. explosion limits
              .. ignitability

QUESTIONS:

2.)   What information on the process would be relevant?

3.)   Having compiled all this information, do you feel able to solve the problem at your desk?

ANSWERS:

2.)   At least, you should have information on:

   · quantity of substance involved per batch/unit operation
   · type of operations, temperature, pressure
   · safety measures already applied
   · maximum amount of substance within one container/unit

3.)   You could only think of one - probably not realistic - worst-case scenario. (It might even not be the worst one!) In order to get an impression on what could really happen you should have a look at the plant, the process and the people operating it.

ASSUMPTION:

4.)   Assume we have to consider the storage, transportation and distribution of **VINYL ACETATE**.

Vinyl acetate ($C_4H_6O_2$) is a clear colorless liquid. It is used to manufacture adhesives, paints and plastics. It has a flash point of -8°C. Its vapors are irritating to the eyes and respiratory system. If it is exposed to heat or becomes contaminated, it may polymerize. If the polymerization takes place in a container, the container is subject to violent rupture. Vinyl acetate is lighter than water and only slightly soluble in water. Its vapors are heavier than air.
(For more detailed information see attached data sheets.)

Our imaginary facility may mainly consist of a **tank farm**: 4 vertical cylindrical tanks for vinyl acetate, each of them with a capacity of 100 m^3. The tank farm is bunded. Vinyl acetate is delivered by ship or pipeline. From the storage tanks, the liquid is filled into **tank cars** and then distributed all over the country.

QUESTIONS:

5.)   What do you mean would be the worst-case scenario in our facility? Where would it happen?

6.)   Where do you mean would be the highest probability of an accident to happen? What kind of accidents could this be?

ANSWERS:

5.)   The greatest amount of vinyl acetate is concentrated in the tank farm. So, the worst-case scenario should be located there.

Possible events are e.g.:

. leakage of a greater quantity of liquid -> evaporation -> ignition -> explosion and fire.

. polymerization inside a tank due to impurities -> vessel explosion -> domino-effects: failure of neighboring tanks resulting in fire and/or explosion.

6.)   If we assume a good preventive maintenance to exist at our plant, the highest probability of an accident to happen should be expected where most frequently human actions have to be done: this might be the station where vinyl acetate is filled from the tanks into the tank cars. The range of possible accidents will vary from leakages of small amounts of liquid out of sealings, flanges or connections up to complete ruptures of connecting flexible tubings. A possible scenario may be a tank car driving away, still being connected to the transfer station.

Other possible accident scenarios, perhaps not completely in our responsibility, may be traffic accidents of the tank cars.

QUESTION:

7.)   What type of accident shall we select for this exercise?

ANSWER:

7.)   On principle, for analyzing the possible outcomings of the
        selected scenario, it does not matter which accident we select.
        The way of thinking and splitting it up into simple, time
        dependent steps will be the same.

        For the further course of the case study let us look for a
        traffic accident of one of the tank cars. For simplicity we will
        assume that this accident results in the similar consequences as
        the one mentioned above: failure of the greatest valve/pipe
        connected to the tank container itself.

ASSUMPTION:

For simplicity let's assume that there is only our tank car involved in
the accident. For technical details of the tank see attached data sheet.
The accident may have happened at night under moderately overcast
conditions. Temperature of environment and the tank's content is
estimated to be 20°C. There is only a very weak wind of about 1 m/s, the
car has come to rest on a public place or road having an asphalt
pavement. The broken pipe is now situated at the bottom of the tank
container.

QUESTION:

8.)   Into what sequential steps would you split up the following
         events?

ANSWER:

8.)   Although in reality the events will not follow one after another,
         we can divide them into the following steps:

         . outflow (liquid, gas/vapor, two-phase)

         . possibly flashing partially

         . possibly turbulent momentum jet

         . possibly spray release

         . spreading of a pool of liquid (eventually boiling, if its
           atmospheric boiling temperature is far enough below ambient
           temperature, then cooled down to its atmospheric boiling
           temperature)

         . evaporation from a (spreading) pool of liquid

         The following events do not only depend on the scenario assumed,
         but also on the properties of the material released:

         . atmospheric dispersion of a toxic gas or vapor

         . ignition and fire

         . ignition and (partially confined) vapor cloud explosion
           accompanied by secondary effects (overpressure, fragmentation,
           fire, toxic fumes etc)

QUESTION:

9.) Estimate type, rate and duration of release.

ANSWER:

9.)   In this case we have to deal with liquid outflow out of an unpressurized container. So, for a first approximation we can use Bernoulli's law resulting in the well known formula:

$$v = \mu \cdot \sqrt{(2 \cdot g \cdot h)}$$

$\mu = 0.7$ (assumed discharge-coefficient)

$h = 1.4$ m (height of liquid in the container)

$g = 9.81$ m$\cdot$s^{-2} (gravitational constant)

$v = 3.7$ m/s
===========

$V' = v \cdot A$
$V' = 7.2$ l/s

$M' = 6.72$ kg/s
===============

$t_o = M_o/M'$

$t_o = 2080$ s $= 40$ min
=====================

So we can estimate that within less than one hour the container will be completely emptied.

If we had to deal with a compressed gas (instead of our liquid, non-compressed vinyl acetate) we had to use the well known standard equations for gas flow. There, in the first step we had to determine whether the flow is critical (sonic or choked), or sub-critical. (These formulas may be taken from standard books on fluid dynamics.)

Furthermore, if we had to deal with a liquid stored above its atmospheric boiling temperature, there would be a certain amount evaporating immediately (flashing) during or just after the discharge, due to the rest of the liquid being cooled down to its atmospheric boiling temperature.

QUESTIONS:

10.) What might be the final size of the liquid pool?

11.) What might happen next ?

ANSWERS:

10.) The pool size depends upon the existence of limiting walls (bunds), drain-pipes, plainness and smoothness of the ground.

Even public places will never be completely plain. If we would assume an average depth of 1 cm for the liquid pool, we would calculate a surface area of

$$A_1 = V_1/h = 15 \text{ m}^3/0.01 \text{ m} = 1500 \text{ m}^2 = 49 \text{ m} \cdot 49 \text{ m}$$

which seems to be unreasonably high. So, let us assume an average depth of 3 cm for the pool :

$$A_1 = 500 \text{ m}^2 = 22 \text{ m} \cdot 22 \text{ m},$$

which might be possible.

11.) If there is an ignition source, a fire would occur. If our tank car is involved in the fire, and still contains a certain amount of liquid vinyl acetate, the evaporation of its content might result in a vessel rupture, followed by a violent explosion of the discharged vinyl acetate cloud. Even if the tank car is not directly involved in the fire, it might "explode" due to the heating of its content by heat radiation. Furthermore, the discharged liquid may polymerize due to contact with impurities or due to exposure to the heat of the flames.

If no ignition or polymerization occurs, the liquid will evaporate out of the surface of the pool. In this case, we have to consider the atmospheric dispersion of a toxic substance.

QUESTION:

12.)   Let us assume that no ignition occurs. What might be the natural
         rate of evaporation ?

ANSWER:

12.)   In order to estimate the rate of evaporation, we have to use the
         analogy of heat and mass transfer: The equilibrium concentration
         of vinyl acetate in the surrounding air is given by its vapor
         pressure. On the other hand, to reach this concentration the
         corresponding amount of heat of evaporation is needed. In order
         to get this heat transfer into the liquid, a certain temperature
         difference between the liquid and its surroundings is needed. So,
         if strong evaporation occurs (due to a high vapor pressure at
         ambient temperature), the liquid will begin to cool down as an
         effect of the heat lost by the evaporated material. With
         decreasing temperature also the vapor pressure decreases and the
         rate of evaporation will be reduced. On the other hand, the heat
         transfer from the surroundings into the liquid will be increased
         as an effect of the lower liquid temperature. This, again, will
         accelerate the rate of evaporation. So, we have to calculate the
         equilibrium between heat transfer by conduction from the
         surroundings and heat loss by evaporation. The mathematical
         modelling of these effects yields to a set of equations that have
         to be solved graphically or by iteration.

         Possible sources of heat transfer into the liquid are

         - ambient air (the rate of heat transfer depending on wind
           speed),

         - possible sources of radiation (fire, sun etc),

         - the surface wet with the liquid (as the surface will also cool
           down in the course of time, this process is time dependent).

         Fortunately, the last process is only important in the case of
         liquids having a high vapor pressure (low boiling temperature).
         If the boiling temperature is sufficiently high, the effect of
         the liquid cooling down is of no importance and heat transfer
         from the ground can be neglected.

If mass transfer by the wind is predominant we can use the formula given by SUTTON[1] :

$$m = a \cdot \frac{p_s \cdot M}{R \cdot T} \cdot u^{(2-n)/(2+n)} \cdot r^{(4+n)/(2+n)}$$

Where: p_s = vapor pressure of the liquid

M = molecular weight

R = universal gas constant

T = temperature

u = wind speed

r = radius of the pool

a,n = are related to atmospheric stability

Stability condition:	n	a
unstable	0.2	$3.846 \cdot 10^{-3}$
neutral	0.25	$4.685 \cdot 10^{-3}$
stable	0.3	$5.285 \cdot 10^{-3}$

As we have assumed above that our accident happened at night, under moderately overcast conditions, we have to assume stable atmospheric conditions, too. Using the following data for vinyl acetate and atmospheric conditions,

T = 20°C = 293 K

u = 1 m/s

n = 0.3

$a = 5.285 \cdot 10^{-3}$

M = 86.09 kg/kmol

$R = 8314.41 \text{ J} \cdot \text{kmol}^{-1} \cdot \text{K}^{-1}$

p_s = 0.12 bar (20 °C)

$r = \sqrt{(500 \text{ m}^2/\pi)} = 12.6$ m

we get:

    m = 0.256 kg/s
    ===============

(In case of the 1500 m^2 pool, we had to expect about three times this value: m = 0.716 kg/s.)

QUESTION:

13.) Could you imagine a theoretical model describing the atmospheric dispersion of a toxic gas emitted with the evaporation rate calculated above?

ANSWER:

13.) Unfortunately, the dispersion of pollutants in the real atmosphere can't be described using molecular diffusion theory. On the other hand, dispersion of hazardous and pollutant materials in the atmosphere has been the subject of intense interest for some decades. On principle, we have to distinguish between different types of releases and following dispersion behavior:

TYPES OF RELEASES:

On principle, we have to consider release rates that are varying with time. As it is difficult to handle those time dependant source rates, usually only two limiting cases are considered:

Continuous release:

Pollutants emitted at constant rate for a prolonged period of time, such as discharges from vents and stacks. (For accidental releases, this model will overestimate the concentrations at greater distances from the source.)

Instantaneous releases:

These short-time emissions are usually the consequence of the rupture of vessels containing pressurized or liquified gases, the latter resulting in flash evaporation. (For most typical accidental releases, this model will overestimate the concentrations in the near field around the source.)

DISPERSION BEHAVIOR:

Especially in the near field around the source, the behavior of the released material depends on it's density, compared to that of the surrounding atmosphere:

a) **Dense cloud dispersion:**
Gases or vapors that are denser than air have a "negative" buoyancy, in the near field they are spreading under gravitational forces. Once atmospheric turbulence has become the main force of spreading, they behave like neutral clouds.

b) **Neutral cloud dispersion:**
If the pollutant's density is similar to that of air, the so called neutral cloud dispersion is described by the Gaussian plume or puff model for continuous or instantaneous releases, respectively.

c) **Buoyant plume dispersion:**
Clouds which are lighter than air (e.g. combustion gases) will float upwards and are therefore likely to disperse harmlessly.

At greater distances from the source, where the released material is sufficiently diluted by the air, always neutral dispersion will occur. So, in this case study we will only consider neutral cloud dispersion, (due to the relatively slow evaporation, emitted vinyl acetate will be well mixed with air by the wind.)

The basic equations for this case were obtained by PASQUILL[2] both for instantaneous and continuous releases. These formulas are still in use in most of the modern computer programs for atmospheric dispersion calculations.

As most accidental releases are only of limited duration, and in order not to go too much into details and problems of atmospheric dispersion calculations, we will only consider the case of a spontaneous release. (This assumption will hold in those cases, where greater amounts of toxic material are released and concentrations in the order of some ppm are of interest.)

In this case, the cloud will travel along with the wind, and with increasing time and distance it will become more and more diluted. The speed of the travelling cloud is the same as wind velocity. At a given point, directly in wind direction, after a time lag (depending on the the distance from the source and on the wind velocity) the concentration will begin to increase. It reaches it's maximum value when the center of the cloud has reached the point under consideration. When the center has passed the point, the concentration decreases. This time-dependent concentration profile is similar to a Gaussian-distribution around it's maximum value. The maximum groundlevel concentrations, occurring directly in wind direction, due to a spontaneous release that happened also at groundlevel, can be calculated using the formula:

$$c(x,t) = \frac{2 \cdot Q}{\sqrt{(2\pi)^3} \cdot \sigma_x \cdot \sigma_y \cdot \sigma_z} \cdot EXP\{ - \frac{(x - u \cdot t)^2}{2\sigma_x^2} \}$$

where: x = distance in wind direction (m)
t = time (s)
$c(x,t)$ = concentration at point "x" and time "t" (mg/m^3)

Q = total amount of material released (kg)
u = wind speed (m/s)
$\sigma_x, \sigma_y, \sigma_z$ = Gaussian dispersion parameters, describing diffusion in x-(wind), y-(crosswind) z-(vertical) direction (m)

Maximum concentration in a given distance is reached at that time when the center of the cloud just has arrived: $x = u \cdot t$. In this case the exponential expression disappears, the formula reduces to:

$$c_{max}(x) = \frac{2 \cdot Q}{\sqrt{(2\pi)^3} \cdot \sigma_x \cdot \sigma_y \cdot \sigma_z}$$

The σ-parameters are necessary to adapt the Gaussian model to the behavior of the real atmosphere. To do this, atmospheric stability is divided into different diffusion categories, depending on solar radiation and wind speed. The parameters are usually given in the form

$$\sigma_y = a \cdot x^b$$
$$\sigma_z = c \cdot x^d$$
$$\sigma_x = \sigma_y .$$

---

[2] F.Pasquill: The Estimation of the Dispersion of Airborne Materials, Met. Mag (1961), 901063,33

For our accident scenario, as described above, the following values were taken from the VDI-guideline 3783 part 1[3) :

    a = 0.6511
    b = 0.785
    c = 0.7544
    d = 0.702

Introducing these parameters into the formula and working it out, we are able to calculate the resulting maximum concentrations to be expected from our hypothetical accident at different distances. However, we always have to bare in mind that these results only are valid for the meteorological conditions assumed (see assumption before question 8).

QUESTIONS:

14.)   In order to do these calculations, we first have to estimate the total amount of vinyl acetate evaporated, depending on the time available for evaporation. How long would you estimate the time for an intervention to stop evaporation?

15.)   Let's focus only on the problem of acute toxicity: Up to what distances should we expect dangerous concentrations to occur? Should we plan for an emergency evacuation?

ANSWERS:

14.)   This answer really depends on local circumstances:
Who would realize that the accident did happen, call the fire brigade and what actions would be undertaken? If nothing happened, evaporation would go on until the total amount of liquid is consumed, ... or if the liquid would disappear through drainage openings into the waist water system.

For this case study let's assume that after half an hour evaporation is stopped by appropriate means.

15.)   If the evaporation only lasts for 30 min, the total amount evaporated is:

$Q = m \cdot t = 1800 \ s \cdot 0.2557 \ kg/s = 460 \ kg.$

Introducing this value into the above formula, we get the following results:

[3) Guidline 3783 Part 1: "Dispersion of Pollutants in the Atmosphere, Dispersion of Emissions by Accidental Releases - Safety Stude", May 1987, Beuth Verlag GmbH, Berlin und Köln

Distance (m)	Concentration (mg/m^3)
300	431
400	224
500	135
600	89
800	46
1000	28
1500	12
2000	5.8
2500	3.5
3000	2.3
3500	1.6
4000	1.2

From this table we see that at distances greater than 600 m the maximum concentrations are below the STEL-value $(60 \ mg/m^3)$, at 800 m the TWA-value is reached, but values below the air odor threshold value only occur at distances of more than 3 km.

It is often difficult to decide whether searching safe shelter in a house, closing doors and windows, or evacuation should be preferred. If, e. g., in our case we assume the STEL-value to be a critical concentration for acute toxicity (which really is not the case), higher concentrations would occur only at distances below 600 m. However, the time for the cloud to reach that distance is about 10 min. So, there would be no time to prepare an evacuation.

Assumed Tank Car

Container-capacity: 20 m³
Container-diameter: 2 m
Contents: 15 m³ = 14000 kg
Liquid-level: 1.4 m
Filling-valve: 5 cm (diam.)

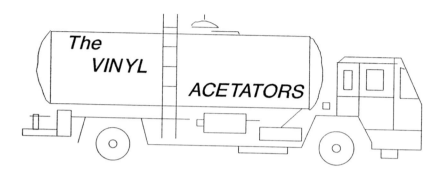

DATA:
Vinylacetate, Acetic acid ethenyl ester,
Acetic acid vinyl ester, Acetoxyethylene, Ethenylacetate

CH3COOCH=CH2
liquid at room temperature

colour	: colourless
molecular weight	: 86.09 g/mol
flash point	: -8 °C
LEL	: 2.6%
	90 g/m3
UEL	: 13.4%
	480 g/m3
auto ignition temp.	: 385 °C
melting point	: -93.2 °C
boiling point	: 72 °C
vapour pressure	: 123 mmHg(20 °C)
	: 164 mbar(20 °C)
specific gravity-liquid	: 0.932 g/cm3 (20 °C)
TLV -time weighted average	: 10 ppm
TLV -short term exposure limit: 20 ppm	
IDLH	: 151 ppm
	: 540 mg/m3
solubility	: 20'000 ppm (25 °C)
vapour density	: 3.0 (air = 1)
odour detectable at	: 0.12 - 0.4 ppm
extinquishing method	: carbon dioxide, dry chemical, "alcohol" foam waterspray may be ineffecive
reaction, explosion with	: Dibenzoylperoxide, Hydrogenperoxide Polymerizationinitiators
exothermal reaction with	: Aluminiumoxide, strong bases, air, strong acids, Toluene,

Panel 9: Inertisation

MODERATORS
T. Fannelop and A. Aellig

CASE STUDY INERTISATION

Flammable solvents and explosive dusts are often used today in the chemical industry. The explosion risks which are linked to the application of those substances are widely known. One of the most effective protection measures is the so called inertisation. By that is meant the displacement of the oxygen which is contained in the air by inert gas so as to make an explosion impossible. This is mainly a fluid-dynamic process of gas streams in a more or less complex shaped apparatus. But also some knowledge of thermodynamics and of the explosion properties of dusts and flammable vapors is needed.

Even though this safety measure is widely used in industrial practice we do not find the appropriate courses at technical universities. The effective inerting of all kinds of apparatuses such as reactors, driers, mills, conveyors, mixers, centrifuges etc. is not so easy and needs very deliberated work. The goal is to get a high degree of safety in production without using a large quantity of inert gas. This requires theoretical knowledge and some practical experience. The latter has to be gained in industry, but the theoretical knowledge must be obtained at universities. This requires a good basic education in this particular field which still is not included in any curriculum today.

The following case study will present some of the problems we may encounter if we want to inert a centrifuge.

DESCRIPTION OF THE STARTING SITUATION

5000 l of a suspension of a pharmaceutical active agent in Hexane should be centrifuged at room temperature (20°C) in several batches. A pendulum centrifuge with a basket of 1000 mm diameter and a maximum speed of 600 revolutions per/min. is available. After each batch the filter cake is being carried out manually. The centrifuge is connected to the building's exhaust air system. The exhaust air is being directed outside through an exhaust air treatment plant. Additionally the following data are given *):

- Material data of hexane

  - Vapor pressure at 20°C: 159 mbar abs
  - Boiling temperature: 69°C at 1013 mbar abs
  - Density of fluid at 20°C: 0.659 g/ml
  - Relative density of the air which has been saturated with vapor at 20°C (air = 1): 1.33

---

*) - Sicherheitstechnische Kennzahlen von Flüssigkeiten und Gasen, SUVA, 1989, CH-6000 Luzern
   - Explosionsschutz-Richtlinien, Berufsgenossenschaft der Chemischen Industrie, 1989, obtainable from Druckerei Winter, Postfach 1061 40, 6900 Heidelberg 1

● Safety data of hexane
  - Explosion limits: 1.15 - 6.3 Vol.% (42 - 242 g/m^3)
  - Flash point: -22°C
  - Range of ignition: -29 - 0.5°C
  - Inflammation temperature: 233°C
  - Minimum limit of O_2 concentration when inerting with N_2: 12.1%
  - Minimum limit of O_2 concentration when inerting with CO_2: 14.5%

● Drawing of the centrifuge before the installation of the inerting:

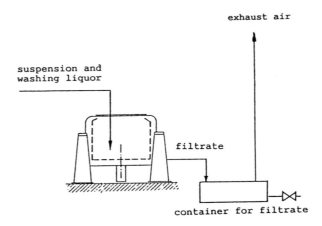

General comments to the following answers:

We have tried to formulate into answers the experiences which have been
gained in practical operation. These answers are a reflection of the
solutions to some local problems in some sites. They do not claim to be
complete and are also not the only solutions, since there are different
optimal solutions depending on the local conditions.

**1a) How large approximately is the maximum concentration of hexane in the
      centrifuge?**

Answer:

The vapor pressure of the fluid determines the maximum concentration. Hence
we obtain:

$$C_{max} = 159 \text{ mbar}/1013 \text{ mbar} = 0.1569 = 15.7\%$$

**1b) The concentration which can be expected in practice, is lower than the
      maximum possible one. By what is it influenced and up to what level do
      you expect it to rise?**

Answer:

The concentration is being influenced by:

- temperature of the hexane

- quantity of penetrated air through leakage. The necessary sub-atmos-
   pheric pressure is produced by ventilation action of the centrifuge and
   by sub-atomospheric pressure in the building-ventilation.

   We expect a concentration of approximatly 5-15%, as there is an almost
   complete saturation due to high turbulences within the centrifuge and
   because there is only a small amount of air, which is being sucked in
   due to leaks.

1c) Is the mixture explosive?

Answer:

The mixture is slightly below or above the upper explosion limit, however, the explosive range is certainly passed through at the start of operation. In any case, we have to take into account that the mixture is explosive at least some of the time.

2) Is there any kind of danger for the employees, the building, the surrounding area or something else? If so, what kind and what is the reason?

Answer:

- Yes, the centrifuge could explode, since the atmosphere is explosive, at least at each start, and several ignition sources are possible.

- If combustible materials or fluids are in the vicinity of the centrifuge and can be ignited by the explosion, an actual danger exists for the building and the surrounding area.

3) What could possible ignition sources be?

Answer:

- Electrostatic charging (hexane is especially dangerous in that case)

- Friction of one metal part on another or heating up of some seal, and hence locally producing temperatures above the inflammation temperature (which is relatively low for hexane).

- Electrical installations within the centrifuge lead to sparking (i.e. insulation failure), if the right protection measure not has been selected. (Zone 0).

4) Which protective measures are principally available for the centrifuge?

Answer:

- Avoidance of potentially explosive mixtures by using non-combustible solvents or by choosing a solvent with a flash point below the working-temperature.

- If possible, lowering of the suspension temperature below the flash point.

- Explosion pressure relief is at least theoretically possible. In practice, however, there are many difficulties which make a realization practically impossible.

- Displacing of atmospheric oxygen by a suitable inert gas (= inerting), so that an explosion is no longer possible.

=> In order to continue the case study, we assume that as a protective measure inerting has been chosen.

5a) **Which inert gases could be used?**
    What are their advantages and disadvantages?

Answer:

	Advantages	Disadvantages
N_2	. extensively inert . availability and normally favorable price depending on site/country	. rather lighter than air . somewhat larger demand for inert gas . danger of suffocating with large leaks
CO_2:	. good inerting effect . heavier than air . easily produced	. chemically not inert to many substances . is well absorbed and dissolved to a great extent . danger of suffocating with small leaks
H_2O:	. very low-priced . plentiful everywhere . inerting effect . similar to N_2 . no danger of suffocation	. condenses at temperatures below 100°C and therefore often only insufficient concentrations attainable
Ar	. heavier than air	. very expensive
He		. extremely lighter than air . very expensive

5b) **Which inert gas would you choose and for what reason?**

Answer:

For the centrifuge we select N_2, as no reaction with a pharmaceutical product can be tolerated.

=> In order to continue the case study we assume that N_2 has been chosen as the most suited inert gas.

6) **If an explosion within the centrifuge should be made impossible, independent of the hexane-concentration (which corresponds to the goal of the so-called partial inerting), how much O_2 can still be tolerated a) in theory?   b) in practice?**

Answer:

a) The highest tolerable O_2-concentration for excluding an explosion depends on the solvent (combustible gas) which is responsible for the potential explosion, and also on the inert gas which is being used. For hexane, inerted with N_2, this comes to 12.1% O_2. If it is not known which solvents are detrimental or in the case where several different ones have to be put to use, then one can assume that the deepest minimum limit of O_2 concentration for still getting an explosion of practically all combustible solvents come up to 10% O_2 if inerted with N_2. (Exceptions are, for instance, combustible gases such as CO with 5.4% and H_2 with 5% O_2.)

b) One should keep in mind that in practice the measurements are generally not 100% accurate. For this reason it is recommended to introduce a safety margin which depends on the reliability of the personnel and the accuracy of the measuring instrument used. For our case we recommend a safety margin of 2%. With this the process should be inerted so as not to exceed 8% O_2.

7) The following task helps us to recall how to use the ternary-diagram. The following two diagrams show us the same process: In a container filled with air (corresponds to the point A with coordinates 21% O_2, 79% N_2, 0% solvent vapors) a solvent is injected, the vapor pressure of which is 0.6 bar abs. The temperature remains constant. After some time the air is being saturated with solvent vapors. The final concentration of the solvent will be 60% (point B). In the diagram on the left the point B has the coordinates 8.4% O_2, 31.5% N_2, 60% solvent vapor. In the diagram on the right, point B has the coordinates 0% N_2 (to be understood as additional inerting gas), 40% air, 60% solvent vapor. Try to understand the diagrams.

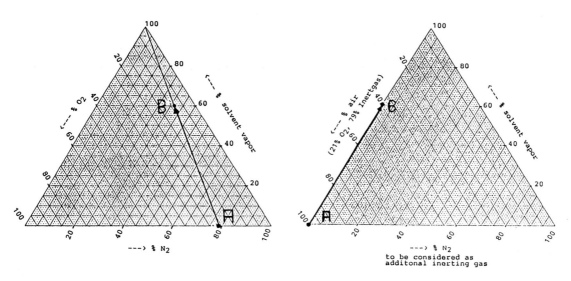

8) In the following ternary diagram "hexane-air-N_2" the final state of the following process-steps should be indicated as points:

   a) Discharging of the filter cake, closing of the centrifuge (it is still moist from the hexane). The air within the centrifuge is being saturated with the hexane. The inerting has not started yet. => point A

   b) Inerting of the centrifuge with N_2 (containing 1% O_2 corresponds approximately to 5% air) up to an O_2-concentration of 5% (which corresponds to 5/21* 100 = 23.8% air) ==> point B. We assume that the process will tanke place very quickly so that during this time period no hexane re-evaporates.

   c) Re-evaporation of solvents from the liquid up to the saturation of the atmosphere within the centrifuge ==> point C (corresponds with the "working point" of the inerted centrifuge).

   d) Mixing of the gases which come out of the centrifuge (i.e. exhaust air) with the surrounding air up to a concentration of 0.5% solvents within the air (distinctly below the lower explosion limit. => point D

Answer:

See drawing

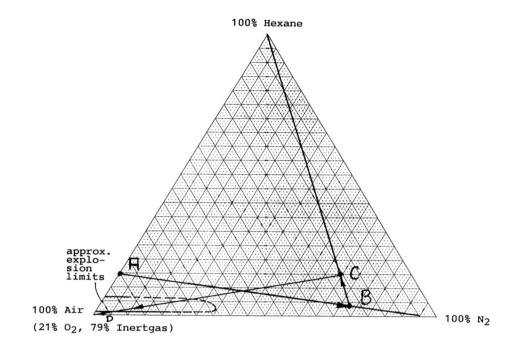

100% Hexane

approx.
explo-
sion
limits

100% Air

(21% O_2, 79% Inertgas)

100% N_2

9) Is total inerting possible? That is, is it possible to inert the centrifuge such as to prevent the leaking gases from forming an explosive mixture with the air outside the centrifuge once they leak out. It may be the case that you have to inert down to a deeper rest of oxygen-content. If so, to how much O_2?

Answer:

In the answer of the question 8d the line from point C to D represents all the different concentrations during the dilution process with ambient air. Now the question has to be asked: what is the criterion for total inerting? In order to be able to speak of total inerting, the straight line, which represents the conditions of a dilution with air, depicted in the ternary diagram, must not cross the explosion limits. It is thereby guaranteed that at no time an explosion occurs during the period of continuous dilution. In an extreme case, the straight line representing the dilution could be tangent t to the field of the explosive concentrations.

From the diagram we see that in our case, total inerting is not possible, even if inerting goes as far as 0% O_2, since all the operating points in the centrifuge are above the tangend t.

In practice, the exact explosion limits are known for few solvents only. For this reason, in many cases the indication of the lower explosion limit and general information about the minimal O_2-limit concentration of 10% for solvents and combustible gases will have to suffice. The zone of explosion concentrations is hence simplified to a triangle. The three boundary lines are the concentration lines 0% N_2, 10% O_2 (/= 47.6% air), 1.15% hexane (lower explosion limit). Although it is slightly larger, this however does not represent a safety problem. In order to evaluate whether total inerting is possible, we draw a line from the point 100% air through the corner of the triangle, point 10% oxygen and 1.15% hexane. Then we evaluate if the operating point is above or below that line. Total inerting ca. be desirable for our centrifuge, but because of the small volume, it is definitely not a requisite since the centrifuge is built in a room which is designated as "zone 2".

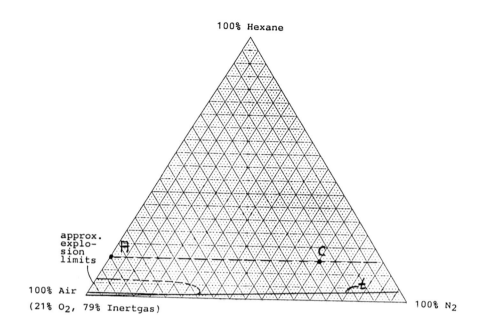

10) Close the centrifuge and let N_2 carefully stream in the switched-off centrifuge the air of which has been saturated more or less with hexane:

a) Does one have to consider the formation of possible gas-layers?

b) Where would one find the N_2 and which results would that bring for the inerting?

c) Do you think that special measures should be taken? If so, specify.

Answer:

a) A distinct lamination has to be taken into account, as the air, which has been saturated with hexane, is 1.33 times heavier than normal air.

b) It is unlikely that the N_2 would even get into the basket, as it would flow down outside the basket. This would happen because the basket would be full of the heavy gas-mixture. The nitrogen would flow from the space between casing and basket into the filtrate-receiver and from there would go into the exhaust air. The result would be that the outer space would be well inerted, but the basket not at all. At the start of the operation everything would get mixed and hence the O_2-concentration would be essentially above the tolerable limit.

c) Yes, the speed by which the N_2 streams in has to be increased so as to obtain a sufficient turbulence which allows adequate mixing. This is possible if we choose as small N_2-inlet diameter in order to get a high speed flow of for example 30-50 m/s).

11) **Please think about which basic concept you would propose for the inerting process. Possible goals would be:**

  - Guaranteed under-pressure (with corresponding measuring) within the centrifuge, so that no solvent vapors can escape into the surrounding area.

  - Guaranteed overpressure (with corresponding measuring) within the centrifuge, so that no O_2 is being sucked in.

  - Continuous, sporadic or no measuring at all of the O_2 concentration (taking into consideration the possible insufficient reliability of many O_2-measuring instruments).

  - Measuring the quantity of the flow rate of the added N_2.

  - etc.

Answer:

Depending on the circumstances, unique to each individual company, different concepts could be the optimal ones. The most common possibilities seem to be:

a) Guaranteed over pressure with alarm, in case the pressure falls. Initial purge with N_2, with adequate control of the N_2-flowrate. Control of the purging time by sporadic O_2-measurements. Addition of N_2 when pressure falls.

b) No over pressure, however continuous O_2-measurements after an initial purging period. N_2-addition with increasing O_2-concentration.

12) **Inerting can be generally carried out completely manually (without help of any automatism), partially automatically, or completely automatically (for example by programmable logic controler).**
Which solution would you choose and why?

Answer:

For an optimal guarantee of safety we choose the automatic operation.

==> For the continuation of our case study we assume the following:

  - An automatic control system has been chosen, in order to guarantee optimum safety.

  - The concept whereby one operates the centrifuge at a constant over pressure has been chosen. First the centrifuge is purged. Then one uses the pressure as a guarantee that no O_2 can enter. The length of the purging-time is chosen in order to be able to guarantee that a concentration of below 8% O_2 is achieved.

13) How should the inerting operation of the centrifuge run? Describe how the inerting should be controlled, how you see the operation procedure point by point, which measuring instruments could be possibly used, and what runs automatically? What has to be handled by the operating staff? Also note the condition which leads to setting off the alarms of the control system. It may help you to do question No. 14 simultaneously.

Answer:

From the many possibilities the following alternative has been chosen with such goals as:

- The inerting process should be affected as little as possible by operating mistakes.

- The inerting process should not depend on the reliability of a O_2-measuring tool.

Concept:

- The centrifuge is constantly kept under a slight N_2 over pressure, so that in case of leaks no air may enter. (The centrifuge serves as a ventilator itself: at the center the pressure can be up to 15 mbar larger than at the outer border depending on the design and rotation speed. So, if the static pressure at the outer border of the centrifuge equals atmospheric pressure because of an open ventilation line. for example, you will find a considerable negative static pressure in the centre of the centrifuge.

- After the operating staff has closed the centrifuge, he starts the inerting at the control panel.

- First the centrifuge is being purged for a certain time with N_2.

- The N_2 escapes by way of an excess pressure relief valve into the exhaust air. The valve opens at an over pressure of approximately 25 mbar.

- After a predetermined time period (established through tests with a O_2-measuring instrument) the purging will stop and the O_2-concentration would have been brought to a desired value of for example 6%.

- Only now is the centrifuge inerted and ready for use.

- Now also the pressure-controller for the pressure in the centrifuge begins to work. If the pressure falls below a limiting value of for example 5 mbar, a valve opens and lets N_2 into the centrifuge until the pressure has reached a nominal value of for example 8 mbar. Then the valve closes again. And the process starts anew.

- In case the pressure falls below a limiting value of for example 3 mbar, an alarm is activated and the centrifuge shuts off automatically.

- In case the N_2-flow is below the given limiting value, an alarm is activated as well (amount of N_2 too small for correct purging).

- Keep in mind: to avoid the lubrication of the bearings to be washed out by solvent-vapors, also the bearings have to be put under N_2-over-pressure (pressure in bearings has to be higher than in centrifuge).

14) Enter into the drawing of the centrifuge:
- where you would like to enter N_2
- where the exhaust air should escape
- how the filtrate should be collected
- where you consider valves, siphons or other parts necessary.

This question you can answer only in connection with question No. 13.

Answer:

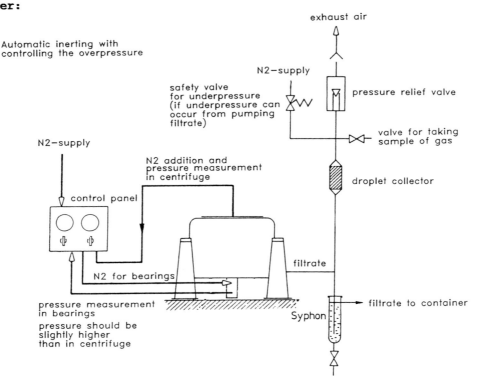

15) In question 10 we have dealt with the problem of producing gas layers, which means bad mixing. But also the form of the centrifuge could, if it is complex or branched out, lead to bad mixing. How can we determine - for instance during the start-up of the plant - whether we have a uniform good inerting within the whole centrifuge.

Answer:

We measure the O_2-concentration during the whole purging period in the exhaust duct. While we continue to measure, we switch the centrifuge on. We make sure that the N_2-supply keeps up during some time. The switching-on of the centrifuge produces a strong turbulence, so that at least by that point good mixing is guaranteed (except for an extreme situation: for example the filtrate container, which is connected to the centrifuge by a duct only, should be hardly effected by the running centrifuge).

If the area of the centrifuge has been well mixed after the purging period, the curve of the O_2-concentration will continue to fall even after the centrifuge has been switched on. In case it has not been well mixed, i.e. there are bad inerted zones, then the curve of the O_2-concentration will show a rise for a short period of time when we switch on the centrifuge. If it is only small, for example from 6 to 7%, then the mixture which occurred during the purging period is acceptable. If it is large however, additional entry-spots for the N_2 are possibly necessary in order to increase the mixing during purging.

Day 4: Summary and the Way Forward

Final panel discussions

MODERATORS
A. Fischli, D. Crowl, S. Cox and V. Pilz

INTRODUCTION

The final panel discussions were chaired by Prof. Dr. J.B. Donnet (CNRS University of Haute Alsace-ENSCM Mulhouse, France). The panel comprised:

Prof. Dr. A. Fischli	COCI, IUPAC, Basel, Switzerland	(Panel 3)
Prof. D. Crowl	Wayne State University, U.S.A.	(Panel 1)
Mrs. S. Cox	University of Loughborough, U.K.	(Panel 2)
Dr. V. Pilz	Bayer AG, Leverkusen, Bayerwerk, FRG.	

Professor Donnet welcomed participants to the final panel discussion and introduced the programme for the session. The moderators of Panels 1, 2 and 3 gave five to ten minutes' feedback on the main issues arising from their separate panels. Dr. Pilz made comments on some of the key issues raised during the workshop and by the separate panels. The moderators of Panels 4-9 gave three minutes' feedback on their case studies (see earlier in proceedings). Dr. Ghosh (Government of Tripura, India) also gave a short presentation on the safety issues in developing countries, together with suggestions for future actions. These inputs were followed by questions from the 'floor'.

This paper summarises the issues raised in the first three panels (including the contribution of Dr. Ghosh to Panel 2) and the open discussions. The case studies (Panels 4-9) are included in the proceedings as valuable educational source material.

PANEL 1 SAFETY EDUCATION AT UNIVERSITIES

Moderators:	D. Crowl	Wayne State University
	T. Kletz	Loughborough University, U.K.

Panel 1 considered the Why? What? and How? of safety education at universities.

Additional comments on safety education in Japanese universities were provided by Professor Uehara.

Why safety education?

Discussions in this area focused on the various pressures (or forces) for safety education at university level including: societal, professional, industrial, financial and, more importantly, pressures from the students themselves. The problems of raising awareness of 'safety' as an academic topic and of influencing faculties to 'major' in safety were pivotal in these discussions.

What safety education?

The Panel conveyed the need to instill motivation for safety in their phrase "Get religion". In addition to such doctrinal and philosophical issues they considered safety technology to be a core topic. Discussions also focused on the requirements of basic educational 'tools', i.e. teaching materials, knowledgeable teaching staff and adequate resources for safety.

How to implement safety education?

A collaborative approach was recommended. Such collaboration should involve liaisons with industry (several existing projects were mentioned) and professional societies. Universities should also network with other universities to pool resources.

Course structures were considered and a mini-course (or short course) approach was favoured in which safety was integrated into the curriculum.

Prof. Yoichi Uehara, Yokohama National University, Japan.

A Comment to "Safety Education at University"

Our department - Safety Engineering, Faculty of Engineering, Yokohama National University, Japan - was established in 1967 and about 800 persons have graduated since then. At the same time, we have taught the lesson of "Outline of Safety Engineering" to the students who belong to the other departments in our Engineering faculty. Lessons concerning chemical safety have been taught to students who belong to industrial chemistry or chemical engineering course in almost all Japanese universities.

From our 20 years' experience, we can say that it is easier to educate our students because they have exact consciousness to the safety and would like to realize a safety society after graduation. But it is difficult to educate the other students because they recognize the importance of safety only vaguely, do not have practical aims and motivations safety. I think how to motivate them to the safety is the most important for the safety education.

Professor Crowl has proposed many items for education at universities (I know he also referred to the importance of motivation). I think they are excellent but it will take much time to realize all of them.

Then I would like to propose a method to realize safety education successfully now. The use of "chemical laboratory" is the required subject for the students of chemistry course. We should improve its content to practice the safety education at universities. We, at first, teach the hazards of materials and reaction processes which are used in experiments. Next, we make students identicate, analyse and evaluate their hazards, and students consider the safety measures to them. I think it is the simplest and most effective way at present.

PANEL 2 SAFETY EDUCATION IN INDUSTRY

Moderators: S. Cox   Loughborough University, U.K.
             G. Clegg   University of Manchester, U.K.

Panel 2 included participants from over 20 different countries worldwide including representatives from the developed and developing countries. The complexity of the topic 'Safety Education in Industry' was reflected in the wide range of comments made by the panel members. The main points are summarised below:

(1) Cultural Differences

The differences in culture, legislation and socio-economic factors influenced the safety educational requirements in various countries. Furthermore, these factors also shaped the size of organisations, the technology and the expectations of the workforce.

(2) Trainees

Safety training was important, not only at the enterprise level but at the level of 'various' organisational groups. This included a 'top down' approach (where training would be required for senior to middle management) and a 'bottom-up approach' covering workers at plant level. The role of trade unions as an educative force was highlighted.

(3) Role of National Governments

The role of national governments in training was discussed. The panel agreed that they needed information both to understand and negotiate technology transfer agreements. They also recommended enforcing authorities to adopt educative rather than policing models of enforcement.

(4) Training Process

Several participants raised the importance of the training process and recommended local training materials (culturally relevant), participative workshops, case study approaches (high trainee involvement) and the use of demonstrations.

Training materials should also include human factors considerations in advanced and developing technologies.

(5) Trainers

The shortage of trainers was seen to be a problem in some countries and various approaches were recommended, including distance-learning materials for individual tuition, training the 'trainers' schools and cascade training methods.

(6) Training in Schools

Panel members recommended that 'safety education' should be carried out in schools (need to educate tomorrow's managers and staff).

Finally, the panel recommended that position papers should be available from each of the participating countries to give an overview of industrial safety training worldwide.

Dr. Ghosh presented a framework of safety in developing countries which included an umbrella approach to legislation and an integrated administration. He stressed the need for institutions to build up the necessary expertise in safety and made the following suggestions for the future:

Suggestions for Future

(1) Develop a system for information storage and exchange.
(2) Institutionalise hazard analysis/safety training arrangement, particularly for managers of small/medium industry on a regional basis.
(3) Frame guidelines for safety, including safety in transfer of technology.
(4) Evolve a mutual industry assistance network channelled through individual IUPAC associated companies in developing countries.
(5) Hold more such workshops.

PANEL 3 FUTURE PRIORITIES IN IUPAC ACTIVITIES ON SAFETY IN CHEMICAL PRODUCTION

Moderators: D. Wyrsch, Ciba Geigy AC, CH
             A. Fischli, Hoffman-La Roche AG, CH

Panel 3 discussed future priorities and considered what IUPAC should do. Should they carry out more workshops? Should they broaden the scope? What should they include in further workshops?

(1) Future Workshops

The panel considered that there should be further workshops and that these should take place fairly rapidly.
        (a) In the first instance this should take place in another industrialised
             country (for example, U.S.A. or Japan).
        (b) In the longer term they recommended a further workshop in a developing
             country to discuss the problems and issues associated with chemical production.

(2) Scope of Workshops

Topics for future workshops should be broadened to include automation/robotics, computer assistance, hygiene factors, environmental issues and storage and handling of chemicals.

Future Issues

Issues for future workshops included:

(a) the provision of model lectures for teaching in industry,
(b) the question of possible guidelines for production in the chemical industry which
     accommodate various legislative requirements,
(c) the issue of entrepreneurial freedom and potential 'safety' restrictions,
(d) the support of centres for safety in the developing countries in association with UNIDO,
     and
(e) safety and biotechnology.

Comment from Dr. Pilz

Towards the end of the panel discussion several speakers pointed out the **importance of transfer of knowledge and experience** in safety technology from experts and more experienced people to others, and some speakers questioned whether the information exchange works even within individual companies.

I agree that the exchange of information is extremely important, and I should like to explain how in Germany we try to guarantee that the necessary transfer of knowledge and experience can take place within companies as well as between different companies.

Within companies we use specialized safety departments to assist designers, plant managers, operators and maintenance crews in solving safety problems. We have also produced guidelines and handbooks on safety and environmental protection, in which we formulate our goals and describe how to proceed in order to meet our goals. ·We have also worked out tools (e.g. checklists) to be used by chemists and engineers in their safety work. Finally we have established standing committees on safety (and environmental protection) for the purpose of exchanging experiences, we publish safety newsletters, and we train and educate people in special courses, seminars, and conferences. Last but not least, the exchange of people by job rotation even with our affiliates in foreign countries plays an important part in bringing safety knowledge and standards up to the same level everywhere.

Outside comapnies the necessary exchange of experience is guaranteed via technical journals, conferences and - most important - via numerous standing committees, where safety experts from
- public authorities
- supervising bodies
- insurance companies
- universities
- engineering and chemical companies

discuss new developments.

I therefore feel that it is virtually impossible to be unaware of any important information on safety issues.

As you all know, at this IUPAC workshop we tried to propagate our knowledge in the developing countries, too.

PARTICIPANTS' QUESTIONS AND COMMENTS

Edward S. Kempa, College of Engineering, Zielona-Gora, Poland.

Question to the IUPAC Organizing and Programme Committee:

Besides detailed discussion on Safety in the Chemical Industry, we should extend our discussions to include, for example, the long-term impacts on the environment. This means combining the feasibility and the availability of technical devices from one side, and the Risk Analysis and Risk Assessment from the other one.

Important examples can be given as e.g. the fall out of the main pumps supplying the water net of a big city or of pumps dewatering the Dutch polders. For how long could the pumps fail? What effects and to what extent are these acceptable?

Ernesto S. Luis, Industrial Technology Development, Department of Science & Technology, Philippines.

Question to D.A. Crowl/S. Cox

What programme or approach would you recommend as being appropriate to implement for education and training of safety in developing countries considering the fact that more than 90% of industries, e.g. Chemical Process Industries, are small to medium scale level?

T.A. Kletz, Loughborough University of Technology, U.K.

Comment:

In transferring technology to less developed countries we should make special efforts to transfer inherently safer and user friendly processes (that is those in which human error or equipment failure has little effect on safety, output on efficiency) for two reasons:
(1) One should transfer our latest and best technology.
(2) Because of lack of experience, equipment failure and human error are likely to be greater in less developed countries.

Other Comment on a Future Conference:

It should include:

More on inherently safer design and user friendly plants.

Demonstration on teaching techniques (slides, videos, discussions).

More on the responsibility of senior managers as shown by analyzing accidents that have occurred.

More discussions of accidents that have occurred.

What can IUPAC do to improve the spread of information between companies on accidents that have occurred? This spread of information is getting worse, not better. If IUPAC wishes to improve the safety record of the Chemical Industry it should try to spread information on accidents by publication, conferences and in other ways.

Luis Puigjaner, Universidad Politechnica de Catalunya, Barcelona, Spain.

Comment:

Plant design and operation is improved at present by adequate computer modelling and simulation that allow for selection of best alternative at lowest cost.

It would be desirable in future workshops to convey information on packages available for computer modelling and simulation of safe design and operation of batch and continuous processes and show the shortcomings, e.g. lack of good dynamic models in the case of batch processes.

B.H. Bibo, AKZO Research, Netherlands.

Remark:

This workshop has been concentrated on all aspects of loss prevention and safety in the Chemical Industry and the emission of normal operations have been discussed in the margin (e.g. industrial hygiene session). It has turned out that there is a lot to improve in the education of chemists and chemical engineers. The initiative for this workshop shows that IUPAC is aware of that problem. However, there is still the problem of chemical products coming in the hands of non chemists and there is hardly any education of the general public on the risks connected with the use of chemical products and the procedures to handle chemicals safely. We will have to do more than labelling and distributing MSDSs. Education of the public is of great importance and IUPAC should take initiatives in this field as well.

Y. Yeannin, IUPAC President, Paris, France.

Comment on the remark of Mr. Bibo:

The future of a chemical in our environment and the use of a chemical by the public is indeed a very important question to be addressed. IUPAC has recognized this need. However, there are so many chemicals that this field is difficult to cover. The commission of Agro-Chemicals has considered the problem for all the chemicals used in agriculture. A workshop about this topic was organized in China. It met a great success. This experiment will be soon renewed in South East Asia.

M. Barreto-Vianna, Alcoa Aluminio, Sao Paulo SP, Brazil.

Contribution:

Organizations like IUPAC could play a more important role in the promotion and dissemination of the transference of technology about risk management safety and industrial hygiene between the industries, universities, government and the society in general.

Special attention must be given to the education and awareness of the public in general (society) what the Chemical Industry is doing regarding the environmental and safety conditions of their operations. More information for the society about environmental and safety subjects is necessary.

Attention must also be given in the risk management transference of technology to decision makers in a more proper way.

Transference of technology on these matters must be made from developed to developing countries, but there are also experiences and opportunities to be learned from developing countries, that could be used in developed countries.

N. Huesch, Winterthur-Versicherung, Switzerland.

Comment:

    Safety = Interdisciplinary in Industry
    Safety = Interdisciplinary at University

- Safety is an interdisciplinary science, involving all engineers and scientists (including mechanical, electrical, civil, architects, physicists, chemists and chemical engineers).

- At universities safety mainly an issue in
  Chemical)
  Nuclear)   individual, non-coordinated activities
  Aviation)
  Civil   )

- IUPAC could take leading role in establishing a safety concept which is uniform and coordinated for all Departments and Faculties at Universities.

R. Rogers, ICI Ltd., Manchester, U.K.

Comment:

This workshop was aimed at Safety Education in Industry and Universities.

As education starts much earlier I suggest one way to improve the situation would be to direct some of IUPAC's effort into ensuring that the concepts of safety including technology/ science, probability/risk and benefits are taught and discussed in the education of children in schools from an early age.

J.H. Shortreed, Institute for Risk Research, University of Waterloo, Canada.

Comment:

IUPAC must contribute to the developing debate on Quantified Risk Assessment versus alternative methodologies (e.g. Zurich approach).

R. Voser, Zurich Versicherungs-Grupee, Switzerland.

Comment:

Public Expectations from University Graduates.

When fixing basic programs for chemists/engineers it may be considered that the public expects that every chemist/engineer has the basic knowledge/understanding on the safety aspects of his field.
e.g. - hazard/risk analysis,
     - basic understanding of hazardous characteristics,
     - basic safety means etc.

G. Kille, Ecole National Superieur de Chimie de Mulhouse, France.

Comment:

1. There are University Departments that have teachings in Safety in Chemical Industry.

2. There are objective limitations within the University working against promoting and developing these teachings.

3. On the other hand, Chemical Industry is anxious that safety education will develop a higher education level and does agree to help.

My proposition is that the Universities that built up such education will join in a Working Party to study this education and contribute to improve it.

L.F. Olivera, Prog. d. Engenhavia Nuclear, Universidade Federal, Rio de Janeiro, Brazil.

Comment:

1. Developing countries should be encouraged to develop their own standards.

2. Universities should offer financial support to young academics to encourage research in safety and risk assessment.

Professor Donnet thanked the conference for their comments and said due account would be taken of them.

Safety in chemical production
Part and parcel of entrepreneurial responsibility

Dr Andreas F. Leuenberger, Vice Chariman
of the Board and Deputy Chairman of the
Executive Committee of Hoffmann-La Roche Inc.
at IUPAC Workshop, Basel, September 13, 1990

Mr Federal Councillor, Ladies and Gentlemen

It is a pleasure for me to join you this morning and speak about 'Safety in Chemical Production, Part and Parcel of Entrepreneurial Responsibility'.

Recently, while reading a well-known Swiss newspaper, I came across an article entitled 'Environmental protection as a corporate goal'. 'Self evident', was my first reaction. But then it occured to me that environmental protection does not - or not yet - hold universal importance as a goal. There are a large number of current examples to prove this e.g. in East European states, where in certain respects, attitudes have remained stuck in the Nineteenth Century. So, environmental protection is an important, but not a traditional goal of industry.

Rather, in the last few years environmental protections has been showing up in the platforms of political parties. Nevertheless, most research-based companies have always taken their responsibilities in this area according to the progress of knowledge. These efforts, however, arise not only from the responsibility a corporation feels towards its employees, the local population and the environment; they are also part of economic considerations which are not limited to a short-term thinking. The integration of safety ecology into corporate policy is a rational process. It helps to ensure the qualitative growth of the organization and thus bring about the achievement of its goals.

The responsibility of the businessman - in fact a compulsory counterpart of his freedom of action - is much broader and no doubt much more enduring than a set of relatively short-term goals. Correctly interpreted entrepreneurial freedom always presupposes acceptance of responsibility for one's own behaviour. On the one hand this includes the duty owed to share-holders because of their capital investment. On the other hand it includes also responsibility for employees' personal safety and job security, responsibility for ensuring an inovative product range and the duty, with respect to our environment, to keep all safety risks well under control. Responsible behaviour of course presupposes strategic thinking aimed at developing and maintaining long-term competitiveness and profit levels.

Safety and environmental consciouness on the part of the corporate manager are therefore a natural part of overall corporate planning. Only by including all relevant factors and considering all possible options can we achieve optimum safety and environmental protection with due heed to market realities. By exploiting modern technology and the possibilities now available for early risk identification to the full, industry is able to act even more quickly and effectively than legislation and regulatory limits require.

This does not mean that industry can constantly achieve the latest technical standards. Investments of some million Swiss francs in order to prevent - for example - air pollution have to be planned carefully. The best available technology has to be evaluated and applied. This process needs time. And therefore, in legislation on environmental protection there is a need of constancy in standards and pollution targets. Any kind of moving targets must be rejected. At the moment of the investment decision, however, industry shows more flexibility. There are many examples in Swiss industry where new investments for safety and for environmental protection not only reached the actual governmental targets but using the available technology even went beyond.

Responsible corporate behaviour is not only idealism. By promoting a durable competitive edge, long-term profits, and sustained corporate health, responsible conduct serves self-interested ends. I am also not forgetting the ethical responsibility of the employer. But it follows from this economic conviction that safety and environmental considerations must be built into corporate policy. The integration of responsible corporate behaviour, finally, into all our activities of our employees is the most effective for minimizing risks for the better use of natural resources.

Responsibility and responsible conduct cannot be expected of the private sector unless the community grants it a measure of autonomy. Without freedom of action, corporate obligations cannot begin to be discharged, opportunities cannot be seized. By that, I do not mean to reject the need for government regulations. Effective government controls are needed in precisely those areas where market forces fail. So far as possible, however, government controls should be in line with free-market mechanisms.

The Swiss Chemicals Industry welcomes the joint establishment of national targets with the state. At the same time Swiss industry has to be allowed the greatest possible autonomy in the choice of the measures to achieve these targets.

Allow me to follow these general remarks by expanding on the notion of responsibility in the private sector. Responsibility, percepted as the complement of freedom of action, manifests itself among other characteristics in self-imposed guide-lines. I have singled out three corporate duties for discussion:

- innovation
- safety
- and new technologies.

Innovation

Innovation is the task of identifying, developing and marketing products and services responding to the needs of today and of the future. This not only requires the establishment of tools within a company searching for novel products - for example the active screening of possibilities offered by the current state of knowledge and technologies. It also requires willingness to take risks. Although the launch of an innovative product to answer a human need has to be carried out in a responsible manner.

Anxious, unconsidered rejection of new technologies prevent any critical and factual assessment and is a reflexion of an attitude which prefers to abstain from solving problems. This attitude can often be observed in spite of clear evidence that new knowledge and technologies and any resulting new products cannot be suppressed. It applies for example to the broad need to develop new pharmaceutical products to provide cures for those two thirds of all identified illnesses which even today remain incurable. Intensive research in the field of genetic information led to the development of gene technology. This for example opened the way to produce a large number of active substances using simple coli bacteria which we would not know for certain which of the hithero incurable diseases will become treatable with the help of this new technology but in view of the tremendous opportunities offered we cannot afford an out-of-hand rejection of gene technology.

Anxiouness is only one of the reasons behind automatic rejection. It can also take place because of incorrect appreciation of the state of knowledge and misjudgement of intellectual capacities. The prudent approach is to form a critical opinion as to the potential risks. This should be done with a determination to find solutions to problems by means of open and factual discussions.

Safety

The IUPAC-Workshop has especially focussed on the safety aspects of production. This involves safety during manufacturing, storage, transport, use and final disposal. It is clearly evident, that in the whole chain of responsibility there are partners of industry as well. All of them, including the consumer, have to assume their share of responsibility.

A fine-tuned and internationally synchronized legal framework guaranteeing enough freedom of action can well serve as a basis for optimal safety in production, handling and disposal of new, innovative products. For the individual user it is essential to acquire the necessary knowledge for appropriate use and disposal.

We are taling primarily about chemical production during these days and the importance of safety in this area cannot be over-emphasized. Entrepreneurship implies change. And change in chemical production must be implemented prudently with a wide ranging assessment which assures that safety is maintained in new or modified products, during the development, construction and start-up process, and in the routine manufacturing of our products. No operation is exempt and no employee is exempt from compliance with and support of safety.

Management's role is not limited to internal activities and efforts but also encompasses corporate responsibilities in the communities in which we live and work and in the physical environment where our products are used and our waste stream disposed. This responsibility includes proactive steps to assure physical safety as well as steps in education to support understanding. As our operations and products become scientifically more complex, the burden of explanation and communication lies with us. Our ability to convey facts and support understanding will assist the public in making informed judgements of product or technology risks and benefits.

Let me at this point comment briefly on what I see as the entrepreneurial opportunities in safety. Our safety experts, and many of you represent that group, must be entrepreneurial and forward thinking. While you deal with the issue and responsibilities of the day, your most important contribution is in anticipation and prevention. It is your responsibility to provide the proper check and balance that assures that safety is an integral part of our products, designed into our plants and a priority component of our education and training programs. This requires constant vigilance and a longer term vision.

There can be no conflict between safety and entrepreneurship. The entrepreneur must always operate within the constraints of prudent safety guidelines. On the other hand, our corporate safety experts must support the internal entrepreneur in his search for new products, new processes and improved efficiency with innovative solutions to the safety issues which surface.

Democratic Society confronted with New Technologies

In democracies, where individuals wield substantial political power, governmental institutions, citizens and industry have to find a common denominator for their interests.

Most established technologies usually find broad acceptance. The scenario is different for new technologies. Large segments of society approach them with substantial doubt and question marks. For the individual, information which enables a personal assessment is not available or is difficult to understand. Experts explaining the benefits and risks act at the same time as opinion moulders. For the businessman the life cycle and potential market penetration of a product or the impact of a technology have been scrutinized and after launch, market progress is only gradually reveiled. For the authorities it is difficult to appreciate the impact of an innovative technology. Even more demanding is the assessment of the influence of a new technology, where no tangible products or services have yet emerged.

We can only enjoy the benefits of innovation if we are willing to develop a common assessment of its impact. So it is important that society improves the way it handles novel products and new technologies. A top priority is open dissemination of what is known about the benefits and risks, which really reaches the consumer.

With regard to the legal framework, internationally well-harmonized legislation could serve to channel ongoing efforts in a desired direction. Scientifically based and internationally accepted standards are certainly a great support for harmonization. Therefore industry welcomes the efforts of non-governmental institutions like the Internaitonal Union of Pure and Applied Chemistry (IUPAC) for their work in setting standards accepted all over the world.

Environmental Protection and Safety in Chemical Production Technical Aspects

In the last twenty years chemical industry made extensive investments in facilities for environmental protection with visible succes. In Germany for instance industrial production has doubled in this period but emissions, arising from production, dropped by 60 to 90 per cent. The share of the total air emissions caused by chemical production in this country is now as low as 3 per cent.

How was this achieved? Chemical production has mainly seen the application of socalled additive environmental protection. In other words, unwanted by-products are separated and eliminated. It will be rather difficult to abandon this procedure completely. The modern approach of integrated environmental protection is to use new production methods supressing the generation of unwanted by-products. New working hypotheses have to be formulated and their technical practicability must be examined in order to find appropriate methods of integrated environmental protection.

An example from polymer chemistry can serve as illustration. Polypropylene was usually obtained by polymerization in a volatile solvent. Minute amounts of the solvent can be detected in the waste air. An elimination of the residual quantities of solvent would have been laborious and expensive. Complete removal has proved to be unattainable. Therefore, additive environmental protection was not the answer.

A new process had to be found. The research target was to find novel catalysis able to induce polymerization without solvents.

Now, modern methods use special catalysts exactly fitting the specific requirements of polypropylene production in absence of a solvent. Searching for integrated environmental protection means finding a tailor-made pathway for each synthesis. There are no broadly applicable general solutions.

When we turn our attention to safety in chemical production a lot of questions could be addressed at and answered during this workshop. If I was asked to pick out one issue from the multitude of topics discussed, my choice would be: minimization of risks. On the technical level this goal can be approached by a systematic reduction of all risk factors, including human mistakes. This can only be achieved by permanent and consequent safety education.

If we ask for enough freedom of action to take the necessary measures, we simultaneously set ourselves an extremely ambitious challenge. Society requires increasingly inter-disciplinary thinking combining economical, technological and ecological considerations. Thus it is first up to us to formulate clear goals for each of these sectors and to achieve them - this is our job. With a better and well-considered risk-management we can demonstrate that we are willing to take risks, but that we also are conscious of our responsibilities and that we accept them.

I thank you for your kind attention.

Struggling for a safe life
The tension between chance and risk

Flavio Cotti, Federal Councillor

Ladies and Gentlemen,

This is a meeting for Chemists, Biologists, Chemical Engineers e.g. for experts in natural and engineering sciences. I am a lawyer and a politician - a non-expert. It is therefore an honour for me to be invited to speak at the first IUPAC (International Union of Pure & Applied Chemistry) Workshop on Safety in Chemical Production.

My situation in front of your community of experts - I have been told from over 40 countries - suggests, that we must have common aims. In order to solve safety problems we should strengthen the cooperation between experts and politicians and improve our understanding of who - experts or politicians - can contribute most to the solution of any particular issue.

1 Hazard, Risk, Safety and standard of living

Our life is characterised by the tension between safety and hazard or chance and risk. This has always been the case, in the middle ages as well as now. The kind of hazard we are confronted with and its level of importance have changed. The life of our ancestors hundreds of years ago was threatened by hazards of natural cause, e.g. weather catastrophes or fire. Technology of man then was fairly primitive, easy to overlook and not very dangerous. The main worries of man were hunger, illness and injuries. Technological progress enabled man to protect himself against natural hazards, the struggle for life became easier. The average duration of the individual life span increased considerably due to the improved standard of hygiene and nutrition. It is interesting to know that from the 13th to the 18th century the average expectation of life at birth in Europe varied between 30 and 45 years. But in the short period between 1870 and 1900 the expectation of life grew to 40 to 50 years. Then we find a continuous increase, and in Switzerland in 1988 the expectation of life at birth was between 74 and 80 years (The variation is based on the fact that women live longer than men).

For the industrialised nations of the western world the hazard concerning life and health has never ever been as small as now. The quality of life based on health and standard of living has improved almost continuously and we find a decreasing tendency for most of the risks in our normal daily life. Consider transport as an example: Although the frequency of using any transport system is increasing, the efforts concerning safety systems and regulation could cope with the situation up to now.

In the public health sector we sometimes find a break-through which can lead to a spectacular reduction of risk. This was the case in the early 50ies when methods of vaccination against diphtheria and poliomyelitis were developed. These hazardous infectious diseases became almost extinct in the industrialised contries after a few years.

In the field of technology the situation is very much different. Individual technical hazards, though they may even be of the same order of magnitude as common natural hazards are not accepted at all. In case such problems turn up unexpectedly, we try to remove them systematically by making use of the technical knowledge in the respective field. In complex technical systems a single new achievement can hardly ever lead to a break-through as it has been mentioned for the medical sector.

Our wealth or prosperity and all human activity are based on the use of technology in the widest sense of the word. Technology today is everywhere, we cannot renounce it, neither in public life nor in private life. From a purely technical point of view it seems possible to realise every conceivable idea. This means that society should evaluate sense and non-sense, value and non-value, profit and loss when applying technical knowledge. Therefore, in order to cope with problems of society now and in the future, we need to consider side by side scientific-technical, ecological as well as human aspects in an adequate combination.

A solution cannot be found - as we are sometimes told - by reducing technology to a "human" level because technology is believed to be the source of all evil. The idea sounds appealing indeed but it is naive. We will need more technology than ever before, it is part of our culture and its implications can never be unravelled. Life without technology is utopia. We therefore must learn to avoid repeating the mistakes of the past in the future. But even by doing so, through extended research and risk assessment, there will always remain a minimum level of uncertainty.

2 Scientific progress, technical progress and safety

Every human being has a desire to feel safe. This is a basic need otherwise we would lose our orientation within the dynamics of society. Also, everybody knows quite clearly that the natural and the civilized environment is inevitably an uncertain neighbourhood. The American inventor Benjamin Franklin said: "Nothing is safe on earth except death and tax-paying".

We are confronted with a constant dilemma. Never ever in our life we can choose or judge something, or take a decision based on complete and perfect knowledge. Our knowledge has always some grade of deficiency, therefore we always take a risk when we decide. In a democratic society this definition of the acceptable risk cannot be left to the industry concerned nor to a few experts. It has to be evolved (and adapted to changed circumstances) in a continuous dialogue between science, industry, public authorities and the public.

It is important to know that man has never been so conscious about risk-taking as now. Because of the elimination and reduction of important hazards, a great number of small ones became more obvious. The fact that we achieved such a high standard of living causes fears that are threatening. The requirement for safety has become extremely important. The job of a workman in a factory must not entail a hazard for his health. Transport of chemicals on the road or on rails must be safe, that means, it must be based on low risk probability. Chemistry is one of the first disciplines and sectors in industry where the discussion about safety played an important role. Biotechnology, molecular biology, and genetic engineering move in the same direction.

As you are experts, there is no need to talk about the content, methods and aims of these disciplines, you know better than I.

It is interesting to note that already at the turn of this century classical biotechnology developed to an important sector of industry. Biotechnology meets basic criteria for a technology of the future: safety, ecological acceptability and innovative potential, and hence comprises elements essential to qualitative growth. On the other hand, the impact of biotechnology - unlike that of traditional technologies - is not "only" of a physical or chemical nature but of a biological one. This new dimension may be the reason, why especially in connection with the development of genetic engineering questions of risk assessment became a major issue in the public over night. The question is not yet answered whether this fact shows a problem of society. Maybe we are confronted with a change in attitude concerning aims of civilisation, or with worries and fear caused by the enormous progress of biology during the past 40 years combined with a lack of information of the public. Here, I mean especially generally conceivable information. In central and northern countries of Europe we find a lot of opposition against genetic engineering whereas in the USA, Japan and southern Europe we find generally neutral or positive attitudes.

The character of modern society with its continuously more exacting demands and, paradoxically, its decreasing readiness to assume risks may explain its attitude towards molecular biology and particularly towards genetic engineering as an applied discipline. Since the potential of genetic engineering became roughly discernible early in the eighties, scientific circles too, have begun to study the conjectural risks which might arise from the application of the new knowledge. I am impressed by the scientists who themselves took the initiative and as early as in 1975 at the Asilomar Conference developed a broad discussion on the safety aspects of the new science. A concept was drafted which culminated in a series of guide-lines.

It is understandable that the possibilities of engineering specific changes in the genetic material of a microorganism may be a cause of concern and grave misgivings among the general public as well as among the scientists. Even though it is recognized today that any concrete safety hazard from modified microorganisms would be relatively small, the fact in no way relieves the scientist of his duty to weigh the risk factors carefully and strictly to comply with precautionary safety measures. The safety guide-lines worked out in the USA by the National Institute of Health and the OECD guide-lines which both are generally accepted everywhere, form a useful basis for the implementaion of safety measures.

The hypothetical long-term risks are conjured up by concerns emphasizing possible disturbances of the ecological system and of the natural evolution of the biosphere. It is undeniable that, as the dominant factor of the biosphere, man strongly influences the ecological system. Today this influence has assumed a gigantic global dimension. Genetic engineering is based on discoveries which have been made in nature. It is becoming increasingly clear, that the processes of genetic recombination are manifold in nature itself. The application of these natural mechanisms to practical uses can, like all of man's civilizing functions, affect the environment in which man lives. Although it would appear extremely difficult to predict correctly the probable long-term effects, it is part of man's responsibilities to give intensive thought to the matter. On the other hand biotechnology fulfils a great deal of prerequisites of an ecologically optimized technology. It is based on renewable raw materials and hence integrated in the natural cycles. Therefore a meaningful assessment of risks and benefits can be done only at an interdisciplinary level with the participation not only of biologists, chemists, ecologists and agronomists, but also of experts familiar with the development trends of human society and in a continuous communication process with the public. As an industry of the future, biotechnology can, within limits, help and be a chance in shaping the development of future civilizations - provided that biology as a whole, the fundamental laws governing the mechanism of heredity and the chemical processes in the living cell become part of man's general knowledge and that fundamental ethical principles are respected.

3 Government and economy: the common aims

What can be done in a society that takes the disappearance of small pox for granted and that is terribly shocked about a technical accident which most often is caused by human failure? What can be done to assuage the fears of people who feel that progress in chemistry and biotechnology is inevitably a threat to the environment?

I believe, that there are several approaches. Let me suggest three that could be realistically implemented.

Firstly, improvement of public information is necessary. We must find means of telling the public about the accomplishments of progress e.g. in chemistry and in the new biology. The public must understand that the potential applications of e.g. biotechnology affect many parts of their lives and are not limited to the area of pharmaceutical products and pesticides. In addition realistic assessment about future applications and risks are critical. A further critical element is public confidence, that we are not trading one environmental problem for another. In this context, I think it is extremely important that those working with media accept particular responsibility.

Honesty requires the clear insight, that there <u>is no advantage without risk.</u> We must avoid creating excessive expectations, but at the same time the public should be aware that new microbial products to replace e.g. toxic chemical pesticides are possible if we keep the climate open for innovation.

This leads to my second point. I would like to emphasize the unremittent need for <u>a broad and vigorous base of research.</u> With respect to the discussed scientific topics, research should show a balance between the fundamental work which may eventually lead to commercial products and the fundamental work which enables us to understand the interaction of newly introduced organism with the environment. Risk assessment research and development which presently does not attract many scientists, should be formulated to include basic research, which would make international cooperation in this area possible or easier.

Talking about research, I also want to include education on all levels. I am convinced that college and university level, the education of the students has to be improved by imparting to all of them a better knowledge of technology. Based on such finding, not only the respective experts but all future scientists and engineers would develop a finer feeling for the tension between chance and risk.

Thirdly, I want to mention a few words on regulation. I said earlier on, I am impressed that the scientists themselves started the discussion on regulation in genetic engineering in 1975. <u>Regulation is certainly necessary but the regulatory approach must be balanced,</u> so that we do not repress innovation in research and development. Overregulation leads to reluctance to invest in the future of new industries. Underregulation at the same time leads to lack of confidence by the public and paralysis of the industry, because of public outcry and legal proceedings. With the <u>proposed amendmend to the Swiss federal law relating to the protection of the environment (Umweltschutzgesetz)</u> we believe we have found the right balance. A combination of this balanced approach to regulatory practise, based on current scientific knowledge, combined with the appropriate communication with the public will lead to an exciting future for all sectors of industry not only for those in chemistry or in new biology.

Conclusions

The aim for a safer life can be approached in several ways. I mentioned three topics:

Firstly, at the turn of the 21th century, we live a remarkably safe life. And yet, we are looking for more safety, albeit the standard of living must not to be touched.

Secondly, scientific and technical progress is only possible by facing new risks. An appropriate balance between safety and risk must be found.

Thirdly, government and business both can contribute to relieve the tension between chance and risk. Appropriate means maybe information and knowledge, research and education and balanced regulation.

No one takes a risk voluntarily unless he is convinced to find an advantage which balances the risk. The Prince of Morocco said in Shakespeare's Merchant of Venice: ".... men that hazard all, do it in hope of fair advantages" (Act II, Scene vii).

Thank you very much for your attention.

Closing remarks

PROFESSOR Y P JEANNIN (PRESIDENT, IUPAC)

I wish to thank all those who have contributed in any way to the undoubted success of this workshop - the organising committee, sponsors, speakers, participants, and not forgetting the many demonstrators who have assisted in the preparation, and demonstration, of the case studies.

IUPAC has been proud to be associated with all these efforts and with Safety in all aspects of Production. This workshop has demonstrated the way Universities - so key to the training of the next generation - can closely associate themselves with industry and the developing world for the benefit of all.

This workshop and others are part of IUPAC's engagement for a better environment through better Chemistry. Chemistry and the chemical industry, with its technology and development, are key parts of the solution to the world's environmental problem.

If this workshop is a success - and I believe it to be so - then it must be only the first of many events led by IUPAC. Already we have plans in hand to organise a second workshop - hopefully in Yokohama, Japan in Spring 1993. We hope the lead sponsor will be the Chemical Society of Japan, and I wish them and those involved in the organisation of the workshop all success.

Finally I hope all participants will use all they have learnt here in Basel for the benefit of mankind. I wish you all a safe journey back to your homeland.

IUPAC-WORKSHOP 1990

Total of Countries:	44
Total of Participants:	128
Developing Countries:	25
Participants of Developing Countries:	50
Participants sponsored by the Organizing Committee:	26
Participants sponsored by UNIDO:	12
Total of sponsored Participants:	38
Faculty members (registrated):	36
Faculty members (not registrated):	7
Faculty members total:	43
Sponsors:	36

I U P A C WORKSHOP 1990

Name	Init.	Title	Institute/Company	Dept.	Address	Postcode	City	Country
Aellig	A.	Mr.	Sandoz AG		Postfach	CH-4002	Basel	Switzerland
Altorfer	F.	Mr.	Ciba-Geigy AG	K-24.312	Postfach	CH-4002	Basel	Switzerland
Andersson	R.	Mr.	Eka Nobel AB	Skoghall Kemi	P.O. Box 503	S-663 00	Skoghall	Sweden
Baloyi	R.	Mr.	Dept. of Occupational Health,	Safety and Workers Comp.	P.O. Box 8433		Causeway, Harare	Zimbabwe
Barreto Vianna	M.	Mr.	Alcoa Aluminio S.A.		Av. Maria Coelho Aguiar 215/Bloco C	BR-05804	Sao Paulo SP	Brazil
Bauer	D.	Dr.	Winterthur Insurance	Safety Engineering	Postfach 286	CH-8401	Winterthur	Switzerland
Bergroth	K.	Mr.	Kemira OY	Kokkola Works	P.O. Box 74	SF-67101	Kokkola	Finland
Berkem	M.	Prof.	Marmara Universitesi	Fen-Edebiyat Fak.	Molla Gurani Caddesi/No 16 Aksaray	TR-34456	Findikzade Istanbul	Turkey
Bibo	B.	Prof.	AKZO Research Laboratories		P.O.Box 9300	NL-6800	SB Arnhem	The Netherlands
Bjoerkman	A.B.	Prof.	Instituttet for Kemiindustri	Danmarks Tekniske Hojskole	Bygning 229	DK-2800	Lingby	Denmark
Blair	E.H.	Dr.	The Dow Chemical Company		4 Crescent Court / Midland	USA-	Midland, MI 48640	USA
Bohman	O.	Dr.	Univ. of Uppsala	Dept. Organic Chemistry	P.O.Box 531	S-75121	Uppsala	Sweden
Botha	F.J.	Mr.	Sasol 1		P.O. Box 1	ZA-9570	Sasolburg	South Africa
Breiner	F.	Mr.	Shell International Chemie	Maatschappij B.V.	Oostduinlaan 2	NL-2501	AN The Hague	The Netherlands
Bruestlein	M.	Dr.	Alusuisse-Lonza		Muenchensteinerstr. 38	CH-4002	Basel	Switzerland
Buchmann	H.-D.	Mr.	Bayer AG	ZF-DID Ing.-wiss. Abt.		D-5090	Leverkusen, Bayerwerk	FRG
Buetzer	P.	Dr.			Rebhaldenstr. 2	CH-9450	Altstaetten	Switzerland
Calzia	J.	Mr.	Dir. Securite et Environ.	Rhone-Poulenc	24, av. Jean Jaures/BP 166	F-69151	Decines Charpieu Cedex	France
Carlsson	L.O.	Mr.	Eka Nobel AB			S-445 01	Surte	Sweden
Carmody	T.	Mr.	American Inst. of Chem. Eng.	Dept. CCPS	345 East 47 Street	USA-	New York, N.Y. 10017	USA
Ching	Ch.B.	Dr.	National University	Dept.Chem.Engineering	Kent Ridge	SGP-0511	Singapore	Singapore
Clegg	G.T.	Dr.	Univ. of Manchester	Inst. of Science/Technology	P.O.Box 88	GB-	Manchester M60 1QD	UK
Corigliano	L.	Dr.	Montedipe (ENIMONT Group)		Via Rosellini 15/17	I-20124	Milano	Italy
Cox	S.J.	Mrs.	Loughborough Univ. of Techn.	Centre for Extension	Studies	GB-	Loughborough LE11 3TU	UK
Crowl	D.A.	Prof.	Wayne State Univ.	Dept. Chem. Engineering		USA-	Detroit, MI 48202	USA
De la Mora	R.	Mr.	Instituto Mexicano del Petroleo	Depto. 302	Sue 89 No. 247 EDIF B-1 Col.Cacana	MEX-09080	Mexico D.F.	Mexico
De Voogd	P.	Mr.	Shell Int.Chemie Maatschappij	Dept. CHSE/1	P.O.Box 162	NL-2501	AN The Hague	The Netherlands
de Vries	E.B.	Mr.	Shell International	Chemie Maatschappij	P.O. Box 162	NL-2501	AN The Hague	The Netherlands
Donnet	J.B.	Prof.	Centre National de la	Recherche Scientifique	24, av. du President Kennedy	F-68200	Mulhouse	France

I U P A C WORKSHOP 1990

Name	Init.	Title	Institute/Company	Dept.	Address	Postcode	City	Country
Ehrig	V.	Dr.	Dr.Karl Thomae	Abt.Chem.Produktion	Birkendorfer Str. 65/P.O.Box 1755	D-795	Biberach/Riss 1	FRG
Eigermann	K.	Dr.	Ciba-Geigy AG	Central Safety Service	K 24.3.18 / Postfach	CH-4002	Basel	Switzerland
Engelthaler	Z.A.	Dr.	Research Institute for Ceramics	Chief Exe.UNIDO/CSSR	Vlastina Str. 3	CS-32318	Pilsen	Czechoslovakia
Fannelop	T.	Prof.	ETH-Zentrum	Inst.f.Fluiddynamik	ML H33	CH-8092	Zuerich	Switzerland
Fischli	A.	Prof.	F. Hoffmann-La Roche & Co. AG	Pharma Divison	Grenzacherstr. 124	CH-4002	Basel	Switzerland
Freemantle	M.	Dr.	IUPAC	Bank Court Chambers	2-3 Pound Way / Templars Square	GB-	Cowley Oxford OX4 3YF	UK
Friedrich	V.	Dr.	Institute of Isotopes	Hungarian Academy of Sci.	Konkoly Thege Miklos ut 29-33	H-1121	Budapest	Hungary
Gharbi	S.	Mrs.	Institut Algerien du Petrole		Centre Es-Senia	DZ-3110	Oran	Algeria
Ghosh	S.	Mr.	Government of Tripura	Principal Secretary		IND-	Agartala	India
Gilon	C.	Prof.	The Hebrew University	Dept.Organic Chemistry		IL-91904	Jerusalem	Israel
Glor	M.	Dr.	Ciba-Geigy AG	K 32.3.01	Postfach	CH-4002	Basel	Switzerland
Gowland	R.	Mr.	DOW CHEMICAL EUROPE		Aert van Nesstraat 45/P.O.Box 1310	NL-3000	BH Rotterdam	The Netherlands
Grunder	R.	Dr.	Ciba-Geigy AG		Postfach	CH-4002	Basel	Switzerland
Grzywa	E.J.	Prof.	Industrial Chemistry	Research Institute	8, Rydygiera st	PL-01 793	Warsaw	Poland
Guillemin	M.	Prof.	IUMHT		Rue du Bugnon 19	CH-1005	Lausanne	Switzerland
Gunzenhauser	S.	Dr.	Institut f.Farbenchemie	Universitaet Basel	St.Johannsvorstadt 10	CH-4056	Basel	Switzerland
Gwinner	E.	Mr.	Hoffmann-La Roche AG	Public Relations Group	Bldg. 58/103	CH-4002	Basel	Switzerland
Gygax	R.	Dr.	Ciba-Geigy AG		R-1060.P.18	CH-4002	Basel	Switzerland
Hammer	H.	Dr.	BAYER AG	OC-P/UA - Geb. G8	Postfach	D-5090	Leverkusen, Bayerwerk	FRG
Hastings	D.	Mr.	Chemical Manufacturers Ass.		2501 "M" Street, N.W.	USA-20037	Washington, D.C.	USA
Hawksley	J.	Dr.	ICI	Group Personnel Dept.	Shire Park / Bassemer Road	GB-	Welwyn Garden City AL7 1HD	UK
Haymerle	H.	Dr.	Sandoz Tech. Ltd.		Building 503/455	CH-4002	Basel	Switzerland
Horak	J.	Prof.	Prague Inst.Chemical Techn.	Dept.Organic Technology	Suchbatarova 5	CS-16628	Prague 6	Czechoslovakia
Hyppoenen	P.	Mr.	OTSO Loss of Profits Ins.Co.		Bulevardi 10 B / P.O. Box 126	SF-00120	Helsinki	Finland
Inwyler	Ch.	Mr.			Nebelbachstr. 5	CH-8008	Zuerich	Switzerland
Ivanus	G.	Prof.	Petrochemical Dept.I.IT.P.I.C.		56-58 Caderea Bastiliei St.	R-71139	Bucharest	Romania
Jaecklin	A.	Dr.	Gerling Consulting Gruppe		Dufourstr. 46 / Postfach 170	CH-8034	Zuerich	Switzerland
Jeannin	Y.	Prof.	Universite P.et M.Curie		4, Place Jussieu	F-75252	Paris Cedex 05	France
Jernqvist	A.	Prof.	University of Lund	Chemical Centre	P.O.B.124	S-22100	Lund	Sweden

I U P A C WORKSHOP 1990

Name	Init.	Title	Institute/Company	Dept.	Address	Postcode	City	Country
Jinghua	W.	Prof.	Inst. of Geography	Chinese Academy of Sciences	Building 917, Beishatan		Beijing 100101	China
Kavala	M.	Dr.	Vyskumny Ustav Pre	Petrochemistry	ul. kpt. Nalepku II c. 86	CS-97104	Prievidza	Czechoslovakia
Kempa	E.S.	Prof.	Institute of Sanitary Engin.	College of Engineering	Podgorna 50	PL-65-246	Zielona-Gora	Poland
Khattab	M.L.	Mr.	Haifa Chemicals Ltd.		9-Moustafa Kamel, Semouha P.O.B.1809	ET-	Alexandria	Egypt
Khermoush	E.	Mr.	Haifa Chemicals Ltd.			IL-31018	Haifa	Israel
Khor	E.	Dr.	University of Liverpool	Inst. of Medical and	Dental Bioengineering/P.O. Box 147	GB-	Liverpool L69 3BX	UK
Kille	G.	Dr.	Ecole Nat.Sup.de Chimie		3, Rue Alfred Werner	F-68093	Mulhouse	France
Kletz	T.A.	Mr.	Loughborough University of	Technology	64 Twining Brook Road/Cheadle Hulme	GB-	Cheadle Cheshire SK8 5RJ	UK
Kroeger	W.	Prof.	Paul Scherrer Institut			CH-5232	Villigen (PSI)	Switzerland
Kuenzi	H.	Dr.	F. Hoffmann-La Roche & Co. AG	Abt. Sicherheit und Umweltschutz	Postfach	CH-4002	Basel	Switzerland
Kurmann	J.	Mr.	Cap Gemini (Schweiz) AG		Grosspeterstr. 23	CH-4052	Basel	Switzerland
Lemkowitz	S.	Dr.	Delft University of Technology	Dept.Chemical Engineering	Julianalaan 136	NL-2628	BL Delft	The Netherlands
Lim	T.T.	Mr.	Inst.Kimia Malaysia	Dept. of Chemistry	Jalan Sultan/Selangor Darul Ehsan	MAL-46661	Petaling Jaya	Malaysia
Lin	S.P.	Mr.	Chinese Petroleum Corp.	Research Institute	83 Chung Hwa Rd. Sect.1	RC-10031	Taipei	Taiwan R.O.C.
Lin	W.S.M.	Mrs.	Industrial Technology	Research Institute	195 Chung Hsing Rd./Sec. 4 Chutung	RC-31015	Hsin-chu	Taiwan R.O.C.
Liu	X.	Prof.	Beijing Inst. of Chemical	Technology	Box 23		Beijing 100029	China
Locher	R.	Mr.	Reto Locher & Partner AG		Elisabethenstr. 44	CH-4051	Basel	Switzerland
Londono Velez	J.	Mr.	Andercol S.A.		Apartado 2065	CO-	Medellin	Colombia
Lowrance	W.	Prof.	Rockefeller University	Life Science/Publ.Pol.Prog.	1230 York Ave	USA-	New York, NY 10021-6399	USA
Luis	E.S.	Dr.	Industrial Techn. Develop. Inst.	Deputy Director, R+D	P.O. Box 774, Ermitta	PI-	Manila	Philippines
Luzny	E.	Mr.	Zaklady Azotowe		ul. Kwiatkowskiego 8	PL-33-101	Tarnow	Poland
Marcano	M.	Dr.	Escuela de Quimica	Facultad de Ciencias	Universidad Central de Venezuela	YV-1040	Caracas	Venezuela
Martinez de la Cuesta P.	P.	Prof.	University of Malaga	Dept.of Chem.Engineering		E-29071	Malaga	Spain
Mavichak	V.	Director	Office of Hazardous Substances		57, Prasumane Road	TH-10200	Bangok	Thailand
Meder	H.	Dr.			Neubadrain 12	CH-4102	Binningen	Switzerland
Mendez	B.	Dr.	Escuela de Quimica	Facultad de Ciencias	Uni.Central de Venezuela/Apt.de Correos	YV-	Caracas	Venezuela

I U P A C WORKSHOP 1990

Name	Init.	Title	Institute/Company	Dept.	Address	Postcode	City	Country
Mendez	B.				47102			
Merz	A.	Prof.	University Regensburg	Inst.Organic Chemistry	Universitaetsstr. 31	D-8400	Regensburg	FRG
Miani	A.	Dr.	ENIMONT		Piazza della Repubblica 14/16	I-20124	Milano	Italy
Moravek	P.	Prof.	Slovak Techn. University	Faculty of Chem. Techno.	Radlinskeho 9	CS-812 37	Bratislava	Czechoslovakia
Muller	St.	Mr.	Audit Sec. des Procedes	Rhone-Poulenc	Interservice-Dose-CP 106	F-69266	Lyon Cedex 09	France
Mussini	F.	Prof.	University of Milan	Dept.of Phys.+Chem.	Via Golgi 19	I-20133	Milano	Italy
Niess	T.	Dr.	Chemie AG Bitterfeld-Wolfen			DDR-4400	Bitterfeld	East Germany
Nishikawa	R.	Mr.	Japan Chemical Industry Ass.		Tokyo Club Bldg. 2-6, 3-chome Kasumigaseki, Chiyoda-ku	J-	Tokyo 100	Japan
Nueesch	K.	Mr.	Winterthur-Versicherung	Int. Div./Saf. Engineering	Rudolfstr. 1	CH-8401	Winterthur	Switzerland
Oliveira	H.J.	Prof.	Prog. d. Engenharia Nuclear	Universidade Federal	Caixa Postal 68509	BR-21945	Rio de Janeiro	Brazil
Onn	L.F.	Dr.	ICI Asia Pacific		P.O.Box 10284	MAL-50708	Kuala Lumpur	Malaysia
Otmanu	A.		ENAD			AL-BP91	El Golea	Algerien
Papp	A.	Prof.	ATOCHEM	Dept. Safety and Environment	Group Elf Aquitaine/La Defense 10	F-92091	Paris 6 La Defense Cedex 42	France
Pardias	R.	Dr.	Ciba-Geigy Ltd.		Esmeralda 2863 / C.P.1602 Florida	RA-	Buenos Aires 602	Argentina
Pasman	J.N.	Dr.	Prins Maurits Lab. TNO	Chem. and Techn. Research Inst.	Lange Kleiweg 137 / P.O. Box 45	NL-2288	AA Rijswijk	The Netherlands
Pavlovic	H.J.	Dr.	University of Belgrade	Fac.of Mining + Geology	Djusina 7	YU-11000	Belgrade	Yugoslavia
Pedelaborde	V.	Mr.	Roussel Uclaf		31, Quai Armand Barbes	F-69250	Neuville sur Saone	France
Peshev	J.	Prof.	Bulgarian Academy of Sciences	Inst.Gen.Inorganic Chemistry	Acad.G.Bonchev Street / Block 11	BG-1040	Sofia	Bulgaria
Petersen	O.M.	Prof.	Technical University	Dept. of Chemical	Engineering	DK-2800	Lingby	Denmark
Pietersen	H.J.	Mr.	TNO	Dept. Industrial Safety	P.O. Box 342	NL-7300 AH	Apeldoorn	The Netherlands
Pilz	Chris	Dr.	Bayer AG	Central Engineering Div.	Building K 9	D-5090	Leverkusen, Bayerwerk	FRG
Pochini	V.	Prof.	Istituto di Chimica	Organica-Universita	Viale delle Scienze	I-43100	Parma PR	Italy
Poeppel	A.	Dr.	Lonza AG	Walliser Werke		CH-3930	Visp	Switzerland
Posniak	K.-L.	Mrs.	Central Institute for	Labour Protection	Tamka 1	PL-00-349	Warsaw	Poland
Puigjaner	M.	Prof.	Uni.Polittecnia de Catalunya	Dept.of Chem.Eng.	E.T.S.E.I.B. Diagona 647	E-08028	Barcelona	Spain
Rademeyer	L.	Mr.	AECI	Engineering Dept.	P.O.Box 796	ZA-1400	Germiston	South Africa
Regenass	D.J.	Dr.	Ciba-Geigy AG	Deputy Director Pigments	Postfach	CH-4002	Basel	Switzerland
	W.							

I U P A C WORKSHOP 1990

Name	Init.	Title	Institute/Company	Dept.	Address	Postcode	City	Country
Reh	L.	Prof.	ETH-Zentrum ML F26	Inst. f. Verfahrens- u. A Prod. Chemikalien	Kaeltetechnik	CH-8092	Zuerich	Switzerland
Reichert	D.	Dr.	Boehringer Ingelheim KG			D-6507	Ingelheim am Rhein	FRG
Richarz	W.	Prof.	ETH	Lab. f. Techn. Chemie	ETH-Zentrum	CH-8006	Zuerich	Switzerland
Rodrigo	F.V.	Mr.	Sandoz Argentina S.A.I.C.		Parana 3617	RA-1640	Martinez-Buenos Aires	Argentina
Rogers	R.	Dr.	ICI Ltd.		Hexagon House, Blackeley/P.O.Box 42	GB-	Manchester M93 DA	UK
Rossinelli	L.R.	Dr.	SUVA		Fluhmattstr. 1	CH-6002	Luzern	Switzerland
Rudolf	H.U.	Mr.	Sandoz Produkte (Schweiz) AG		Postfach	CH-4002	Basel	Switzerland
Rueedi	P.	Dr.	Universitaet Zuerich	Irchel	Winterthurerstr. 190	CH-8057	Zuerich	Switzerland
Russo	G.	Prof.	Universita de Napoli	Dip. Ingegneria Chimica	Piazzale V.Tecchio 80	I-80125	Napoli	Italy
Sacerdoti	S.	Dr.	ABIC Ltd.		P.O. Box 8077	IL-	Natania	Israel
Sakra	T.	Dr.	Inst. of Chem.Technology		nam. Legii 565	CS-53210	Pardubice	Czechoslovakia
Salzmann	J.J.	Dr.	Sandoz Technology Ltd.		Lichtstr. 35	CH-4002	Basel	Switzerland
Samel	U.-R.	Dr.	BASF AG	DUS/A-M 940	Carl-Bosch-Str. 38	D-6700	Ludwigshafen	FRG
Sanner	T.	Prof.	Institute of Cancer Research		Radiumhospitalet Montebello	N-0310	Oslo 3	Norway
Sarmiento	C.R.	Mr.	Tensioactivos Del Litoral S.A.		Parque Industrial / Gualeguaychu	RA-2820	Gualeguaychu Entre Rios	Argentina
Schacke	H.	Dr.	BAYER AG	IN-ATU VA	Geb. E 41	D-509	Leverkusen	FRG
Schaerer	R.	Mr.	Lonza AG			CH-3930	Visp	Switzerland
Schaller	L.	Mr.	E.I. du Pont de Nemours & Co.Inc	N-11466 Dept. Relations	1007 Market Street	USA-	Wilmington, DE 19898	USA
Schiess	M.	Mr.	BUWAL		Hallwylerstr. 4	CH-3003	Bern	Switzerland
Schmalz	F.	Dr.	Ciba-Geigy AG	K 24.3.24	Postfach	CH-4002	Basel	Switzerland
Schneemann	K.	Dr.	Huels AG	ST SI Bau 0139/PB 03	Postfach 1320	D-4370	Marl	FRG
Setzer	A.	Dr.	Colonias Sur No221-Piso 7	Col. Americana	Sector Juarez	MEX-44100	Guadalajara, Jalisco	Mexico
Sharma	S.M.	Dr.	Oil + Natural Gas Commission	Safety/Env.Management		IND-248001	Dehradun	India
Shortreed	J.	Prof.	Institute for Risk Research		University of Waterloo	CDN-	Waterloo N2L 3G1 Ontario	Canada
Siwek	R.	Mr.	Ciba-Geigy AG		Postfach	CH-4002	Basel	Switzerland
Soon	T.K.	Dr.	Tunku Abdul Rahman College	K 32.3.04	P.O.Box 10979/Jalan Genting Kelang	MAL-50932	Kuala Lumpur	Malaysia
Stoessel	F.	Dr.	Ciba-Geigy AG	K 127.5.04	Postfach	CH-4002	Basel	Switzerland
Sugavanam	B.	Mr.	UNIDO Vienna Int.Centre	Dept. Industrial Operations	P.O.Box 300	A-1400	Vienna	Austria
Suokas	J.T.	Prof.Dr.	Technical Research Centre	Safety Engineering Lab.	P.O.Box 656	SF-33101	Tampere	Finland

I U P A C WORKSHOP 1990

Name	Init.	Title	Institute/Company	Dept.	Address	Postcode	City	Country
Suryosunarko	S.A.	Mr.	Director for Agrochemcal Industries		Jl. Tebet Barat No. 21	RI-	Jakarta 12810	Indonesia
Tayim	H.	Prof.	Petroleum & Petrochem. Research Institute		KACST / P.O. Box 6086	11442	Riyadh	Saudi Arabia
Tereshcenko	G.F.	Prof.	State Institute of Applied Chemistry		Dobrolyubov ave 14	SU-197198	Leningrad	USSR
Toth	L.	Dr.	Eszamagyarorszagi Vegyimuvek			H-3792	Sajobabony	Hungary
Tschopp	A.	Dr.	Sandoz Technologie AG		Postfach	CH-4002	Basel	Switzerland
Tuominen	S.	Mr.	Ministry of Trade and Industry		Kluuvikatu 3A	SF-00100	Helsinki	Finland
Uchida	M.	Dr.	Teijin Ltd.		1-1, Uchisaiwai-Cho/2-Chome, Chiyoda-Ku	J-	Tokyo 100	Japan
Uehara	Y.	Prof.	Yokohama National Univ.	Fac. of Engineering	156, Tokiwadai Hodogaya-ku	J-240	Yokohama	Japan
Valenzuela	M.A.	Mr.	Sebatian del Piombo No. 55-B	Int 602	Col. Mixcoac	MEX-03910	Mexico D.F.	Mexico
Venselaar	J.	Dr.	Akzo Engineering BV		P.O. Box 9300	NL-6800	SB Arnhem	The Netherlands
Voser	R.	Mr.	Zuerich Vers.-Gruppe		Mythenquai 2	CH-8002	Zuerich	Switzerland
Vuillard	G.	Mr.	RHONE-POULENC		25 Quai Paul Doumer	F-92408	Courbevoie Cedex	France
Wang	G.L.	Mr.			No.218 Yan An Road 3rd / Shan Dong		Qing Dao 266071	China
Washer	M.	Mr.	Solvay & Cie. S.A.	Direction Centr. Techn.	Rue de Ransbeek 310	B-1410	Waterloo	Belgium
Wegmueller	H.	Dr.	Ciba-Geigy AG	Research and Development	Dyestuffs and Chem. Division	CH-4002	Basel	Switzerland
Welch	G.	Mr.	The Wellcome Foundation		Temple Hill	GB-	Dartford, Kent DA1 5AH	UK
Whiston	J.	Dr.	ICI Group Headquarters	Personnel Department	9 Millbank	GB-	London SW1P 3JF	UK
Widner	A.	Mr.	Chemische Rundschau	Verlag Vogt-Schild	Zichwilerstr. 21	CH-44501	Solothurn	Switzerland
Widner	F.	Prof.	Inst. f.Verfahrens- und Kaeltetechnik ETH		Sonneggstr. 3	CH-8092	Zuerich	Switzerland
Witkowski	W.	Mr.	Chemistry Institute WAT		Kaliskiego	PL-01-489	Warsaw	Poland
Wojtowicz	S.	Mr.	Designing Office of Nitrogen Works		Kedzierzyn	PL-47-220	Kedzierzyn-Kozle	Poland
Wyrsch	D.	Dr.	Ciba-Geigy AG	Dyestuffs and Chem. Div.	K-424,3,18 / Postfach	CH-4002	Basel	Switzerland
Xu	Y.H.	Prof.	Science and Techn.Commission	of Shanghai Municipality	30 Fuzhou Road		Shanghai 200002	China
Yassin	A.	Prof.	University of Cairo	Dep. of Chemistry	Faculty of Science	ET-	Cairo	Egypt
Zaengl	W.	Prof.	ETH-Zentrum	Inst.f.Hochspannungst.	ETH-Zentrum (ETL H 29)	CH-8092	Zuerich	Switzerland

Count: 164

Bibliography of training material for safety in chemical production

IST IUPAC WORKSHOP   Basel, Switzerland - September 9th - 13th 1990

1.   INTRODUCTION

The training of chemical process personnel in safety related areas requires appropriate training materials, databases and supporting literature.   There is a plethora of such items on the market and this bibliography has been compiled to provide useful examples of these.   The author acknowledges that her review is based on her own knowledge and experience of training within the United Kingdom and is not, therefore, totally comprehensive.   A more detailed review could be developed as an ongoing task with the support of this workshop's participants.

2.   TRAINING MATERIALS

2.1   Provider

Chemical Industries Association,
Kings Buildings,
Smith Square,
London SWIP 3JJ.

Materials - **A range of videos, computer based training packages and booklets on chemical safety.**

'Responsible Care', I5 minute video illustrates the origin, philosophy and importance of the UK Responsible Care programme.

'Responsible Care - Responsible Action', twelve months on, it shows not only how the programme has been developed from the centre but also and more importantly, cites practical examples of initiatives taken by companies under the aegis of Responsible Care.   The seven company case studies illustrate all aspects of the programme from improved safety and environmental performance, through employee participation to customer relations and communications with the public.

'Chemicals with Care', this video outlines the special safety measures undertaken by chemical companies to maintain safe conditions both within and around the chemical plant.

'We Care for the Environment', video explains how waste is disposed of or recycled and shows the measures the chemical industry takes to safeguard our environment in general.

'On the Move' this explains the precautions taken by the chemical industry to ensure their safe despatch; including special fittings on tankers, and the container and vehicle markings.   It also looks at driver training, and the support services which are available if a spillage occurs.

'Safe Handling of Chemicals, **'Solving a Process Problem'** and **'Safe Working on Scaffolding'**, these three computer based courses (IBM P.C. compatible) have been produced in conjunction with Process Engineering magazine.   Cost £95 sterling.   Details available from Training Disks, Process Engineering, Morgan-Grampian House, Calderwood Street, London SE18 6QH.

2.2 Provider

Du Pont Company Safety Services
Barley Mill, P19-1210,
Wilmington,
DE 19898

and

Du Pont De Nemours (Belgium)
Industriezone "Mechelen Zuid 1",
Antoon Sponoystraat,
6-2800 Mechelen,
Belgium.

Materials - Du Pont Safety Services offer a broad range of safety training materials for employee training, including:

'Basic Safety Training', a self study course which ensures new employees are aware of basic safety principles.

'Safe Practice Series', includes a leaders manual, sixteen self study workbooks and safe practice series cards, subjects covered include:

Planning for Safety
Back Safety
Confined Space Entry
Flammable Liquid Hazards
Hand Safety
Industrial Housekeeping
Laboratory Safety
Office Safety
Personal Protective Equipment
Pinch Points
Portable Fire Extinguishers
Portable Tool Hazards
Slips, Trips, and Falls
Static Electricity Grounding

'Take Two for Safety', video tape docu-dramas portraying real life scenarios on safety related issues.

'HAZCOM' Reports, meets general training requirements of expanded OSHA Hazard Communication Standard.   HAZCOM Reports programme informs managers and employees about chemical hazards in the workplace and safety compliance strategies.   The format is a combination of workbook and video.

'Managing Safety', safety techniques for professional safety advisers, operations managers and supervisors.

2.3   Provider

Health and Safety Executive (United Kingdom)
Library and Information Services,
St. Hugh's House,
Stanley Precinct,
Bootle,
Merseyside,
L20 3QY.

CFL Vision,
P.O. Box 35,
Wetherby,
West Yorkshire,
LS23 7EX.

Materials - Extensive catalogue of films, videos and slides on general safety including:-

'Danger Contained', this programme outlines the major parts of the UK Road Traffic (Carriage of Dangerous Substances in Packages) Regulations 1986 as they affect manufacturers, fleet operators, hauliers and drivers.   It has been designed to act primarily as an introduction to training in the regulations for enforcers, industry and the emergency services.

'**Deadly Maintenance in the Chemical Industry**', this programme highlights the safety related issues associated with plant maintenance during shut-down.

2.4 Provider

Industrial Training Systems Corporation (USA),
9 East Stow Road,
Marlton,
New Jersey 08053,
USA.

Fax (609) 983-4311

<u>Materials</u> - **General safety videos including:-**

'**What's It All About**', A worker's overview of the COSHH Regulations and occupational hygiene practice.

2.5 Provider

Rank Millbank Training
Cullum House,
North Orbital Road,
Denham,
Middlesex, UB9 5BR.

<u>Materials</u> - **Extensive catalogue of safety films and videos on a variety of topics including:-**

'**Unreasonably Dead**', film concerned with electrical safety of installations.

'**Permit to Work**', sponsored by Oil and Chemical Plant Constructors' Association.

'**It need not happen**', film on protective clothing and equipment

2.6 Provider

Royal Society for the Prevention of Accidents,
ROSPA Film Library,
Cannon House,
The Priory Queensway,
Birmingham B4 6BS.

<u>Materials</u> - **A range of films and videos on general safety including:-**

'**Disaster Planning Exercise Sirocco**' 19 minute video on disaster planning.

2.7 Provider

Technical Video Sales,
Dunley House,
Toft Road,
Knutsford,
Cheshire WA16 9DY.

<u>Materials</u> - **General safety films including videos concerned with safety legislation:**

'**What They Mean To You**', price £450 sterling plus VAT.

'**Electricity Regulations Training Package**', price £575 sterling, plus VAT.

2.8   Provider

The Institution of Chemical Engineers,
165-171 Railway Terrace,
Rugby,
Warwickshire, CV21 3HQ,
England.

<u>Materials</u> - Safety and Loss Prevention, Hazard workshop modules

Information and Slide Training

Module 001 -   Hazards of Over - and Under-Pressuring of Vessels

Module 002 -   Hazards of Plant Modifications

Module 003 -   Fires & Explosions

Module 004 -   Preparation for Maintenance

Module 005 -   Furnace Fires & Explosions

Module 007 -   Work Permit Systems

Module 008 -   Human Error

The above slide training packages cost £190 sterling each to commercial organisations and £80 sterling each to educational establishments.

Module 015 -   HAZOP and HAZAN by Professor Trevor Kletz, price £195 sterling.

Video Training

Module 006 -   Preventing Emergencies in the Process Industries. Price £495 sterling, £250 sterling to educational establishments.

Module 009 -   Inherent Safety. Includes copy of publication "Cheaper, Safer Plants" by Professor Trevor Kletz, price £495 sterling, £250 sterling to educational establishments.

Module 011 -   Safe Handling of LPG - Part 1 - Pressurised Bulk Storage and Road and Rail Loading. Price £750 sterling, £400 sterling to educational establishments.

Module 012 -   Safer Piping. Price £750 sterling, £400 sterling to educational establishments.

Module 013 -   Safe Handling of LPG - Part 2 - Ship-Shore Transfer and Refrigerated Storage of LPG. Price £750 sterling, £400 sterling to educational establishments.

Computer Based Training

Module 010 -   Handling Emergencies

This computer program is the sequel to Module 006, 'Preventing Emergencies in the Process Industries' and it combines both strategic and tactical approaches to the handling of a range of fire and toxic incidents on a process plant in a populated area. Price £750 sterling, £400 sterling to educational establishments.

Open Learning Training

Module 014 -   Practical Risk Assessment

This module enables staff to teach **themselves** to assess the **hazards** of a situation and the associated **risks**.   The workbook teaches a systematic and simple method of assessing and reducing the risks of a job.   The method will help all those issuing work-permits, to identify the risks of routine tasks such as vessel entry, despading a caustic line etc.   Price £195 sterling.

2.9   Provider

Videotel International,
Ramillies House,
1/2 Ramillies Street,
London,
W1V 1DF.

Materials

A video training programme developed with the support of several leading chemical organisations concerned with the Handling and Processing of Hazardous Chemicals.

The subjects of the package are:-

>   The prevention of emergencies,
>   On-site emergency response,
>   Off-site emergency response,
>   Community awareness.

3.   DATABASES

Online and Compact Disc Read Only Memory Services (CD ROMs)

Commercially available online services have proliferated over the past decade making information more easily accessible.   It is estimated that there are well over 4,000 databases available for retrieval in the world covering many subjects and totalling over 100 million references.

The following databases will give useful information:

3.1   Chembank

Coverage: Collection of databanks of potentially hazardous chemicals, RTECS, OMHTADS and TOSCA.   Available from: SilverPlatter Information Services, 10 Barley Mow Passage, Chiswick, London W4 4PH, Tel: 01-995-8242, Telex: 265871, Fax: 01-995-5159.

3.2   Chemical Hazards in Industry

Royal Society of Chemistry,
Burlington House,
Piccadilly,
London WIV OBN

Coverage: All aspects of chemical hazards in industry.   Particularly useful for: chemical and hazard enquiries, including accident prevention, hazardous waste management, UK legislation.   Printed version: Chemical Hazards in Industry - Monthly.

3.3   CISDOC

Centre International d'Information de Securite et d'Hygiene du Travail,
International Labour Office,
Health and Safety Centre,
4 Route des Morillons,
CH-1211,
Geneva 22,
Switzerland.

Coverage: Embraces occupational hygiene, medicine, physiology, industrial toxicology, accident prevention, industrial toxicology, accident prevention, safety engineering. 20,000 references, updated seven times per year.   Particularly useful for: checking worldwide occupational health and safety references, especially legislation.   Printed version: CIS Abstracts. Available on OSHROM CD ROM.

3.4   ECDIN

Environmental Chemicals Data and Information Network,
produced and maintained by:

Commission of the European Communities,
Joint Research Centre,
Att. ECDIN group,
I-21020 Ispra (Varese),
Italy.
Database on chemical substances of environmental importance, particularly useful for: chemical structure, producers plants, health and safety including toxicological surveillance.

3.5   FACTS

Available from:

TNO Division of Technology for Society,
P.O. BOX 342,
7300 AH Apeldoom,
Laan van Westenenk 501,
7334 DT Apeldoom,
The Netherlands.

Coverage: FACTS provides case histories of accidents with hazardous materials which happened worldwide over the last 30 years. It focuses on the following industrial activities: processing,   storage,   transshipment,   transport   and use/application.

3.6   HSELINE (Health and Safety Executive) - see earlier for address

Coverage: All aspects of health and safety associated with chemicals and chemical engineering industry.   Monthly updates.   Particularly useful for: up to date information from world wide sources, UK legislation and incidents information. Printed version: none. Available on OSHROM CD ROM.

3.7   MHIDAS - Major Hazard Incident Data Service information package on the new data bank, available from:

The Library,
Safety and Reliability Directorate (SRD),
Wigshaw Lane,
Culcheth,
Warrington,
WA3 4NE.

Coverage: SRD/HSE reports of major hazard incidents published since January 1989.

3.8   OCCUPATIONAL SAFETY AND HEALTH (NIOSHTICS)

Coverage: bibliographic database produced by the United States National Institute for Occupational Safety and Health.   Articles entered into the database are taken from several sources dating back to 1973 as well as important articles from the early nineteenth century.   Particularly useful for: all aspects of the occupational safety and health field.   Available on OSHROM CD ROM.

3.9   RTECS

Coverage: toxic effects of chemical substances.   US National Institute for Occupational Safety and Health (NIOSH).   40,000 substances, updated quarterly. Particularly useful for: threshold limit values; recommended standards in air and aquatic toxicity levels; toxicity effects. Printed version: Registry of toxic effects of chemical substances. US Government Printing Office, DHEW Pub. Also available in microfiche updated quarterly.

4.   SOURCES OF INFORMATION

4.1   Barbour Microfile Health and Safety

Barbour Microfiles,
New Lodge,
Drift Road,
Windsor,
Berks. SL4 4RQ.

Coverage: the Barbour Health and Safety Microfile has been produced to provide vital reference information for designers, safety officers and others responsible for complying with the requirements of the UK Health and Safety at Work Act 1974. The Microfile contains the full text of publications ranging from legislation to standards and codes, advisory leaflets and data sheets.

4.2   EINECS - European Inventory of Existing Commercial Chemical Substances.

Details available from:

UK EINECS CONTACT Pt (HSD D2),
Baynards House,
1, Chepstow Place,
Westbourne Grove,
London,
W2 4TF.

Coverage: the Commission of the European Communities published the European Inventory of Existing Commercial Chemical Substances (EINECS) in the nine official languages of the Member States on 15 June 1990.   The inventory, which will have full legal effect from 15 December 1990, lists those substances commercially available in the European Community (EC) between 1 January 1971 and 18 September 1981, a total of just over 100,000.   It does not cover medicinal products, narcotics, radioactive substances, foodstuffs or wastes.   ISBN 011 9678969.

4.3   The Society of Industrial Emergency Services Officers (SIESO) - Guide to Emergency Planning.

Available from:

Paramount Publishing Ltd.,
17-21 Shenley Road,
Borehamwood,
Hertfordshire WD6 1RT.

Coverage: a practitioner's guide to cover all types of hazard that could affect industry - natural and man-made, accidental or deliberate.

4.4   HSE Publications

Publications available from HSE Library and Information Services (see earlier)

These publications are listed in 'Publications in Series: List of HSC/E publications January 1990'

4.5   Health and Safety Directory

Published by:

Kluwer Publishing Ltd.,
1 Harlequin Avenue,
Great West Road,
Brentford,
Middlesex, TW8 9EW.

Coverage: this directory is designed to provide a comprehensive reference source for people involved in health and safety at work.   It includes details of training bodies and materials.

4.6   ILO Publications

Catalogue available from:

International Occupational Safety and Health,
Information Centre (CIS),
International Labour Office,
1211 Geneva 22,
Switzerland.

Any request for publications should state the full title, the ISBN and the language desired.

Coverage: publications on occupational safety.

4.7   Loss Prevention Council

Publications available from:

The Loss Prevention Council,
140 Aldersgate Street,
London, EC1A 4HY.

Coverage: comprises publications and visual aids prepared by specialists from the LPC Technical Centre, the Loss Prevention Certification Board (LPCB). the Fire Protection Association (FPA), and the National Supervisory Council for Intruder Alarms (NSCIA).   The list offers a comprehensive range of material designed to assist those whose work demands a professional knowledge of the numerous elements of loss prevention and control.

4.8   National Fire Codes

Codes available from:

National Fire Protection Association,
Batterymarch Park,
Quincy,
MA 02269

Coverage: the National Fire Protection Association was organized in 1896 to promote the science and improve the methods of fire protection and prevention and to obtain and circulate information on these subjects and to help establish proper safeguards against loss of life and property by fire.

5.   RECOMMENDED TEXTS

'Cancer Causing Chemicals' N.I. Sax, Van Nostrand Reinhold, 1981.

'Casarett and Doull's Toxicology: the Basic Science of Poisons' J.D. DOULL, c.d. Klaasen et al, 2nd ed. MacMillan, 1980.

'Croner's Health and Safety at Work' P.A. Chandler and C.T. Stoddard, New Malden, Croner Publications, 1979.

'Dangerous Properties of Industrial Materials' N.I. Sax, B. Feiner and others, 6th ed., Van Nostrand Reinhold, 1984.

'Diseases of Occupation' D. Hunter, 6th ed. Hodder and Stoughton, 1978.

'Encyclopaedia of Occupational Health and Safety' (2 vols). Dr. L. Parmeggiani (ed), 3rd rev. ed, Geneva, ILO, 1983.

'Handbook of Occupational Hygiene' B. Harvey, S. Silk et al, Kluwer Publishing Ltd., 1980.

'Handbook of Reactive Chemical Hazards'L. Bretherick, 3rd ed, Butterworths, 1985.

'Handbook of Toxic and Hazardous Chemicals and Carcinogens' M. Sittig, 2nd ed, New Jersey, Noyes Publications, 1985.

'Hazards in the Chemical Laboratory' L. Bretherick, 3rd ed, RSC, 1981.

'Guidelines for Chemical Process Quantitative Risk Analysis'

Available from: Center for Chemical Process Safety
                  of the American Institute of Chemical Engineers,
                  345 East 47th Street,
                  New York,
                  NY 10017

Other Publications available from the Center:-

'Guidelines for Technical Management of Chemical Process Safety'

'Guidelines for Chemical Process Quantitative Risk Analysis'

'Guidelines for Vapor Release Mitigation'

'Guidelines for Safe Storage and Handling of High Toxic Hazard Materials'

'Guidelines for Use of Vapor Cloud Dispersion Models'

'Workbook of Test Cases for Vapor Cloud Source Dispersion Models'

'Guidelines for Hazard Evaluation Procedures'

'Proceedings of the International Symposium on Runaway Reactions, 1989'

'Proceedings of the International Conference on Vapor Cloud Modeling, 1987'

'Proceedings of the International Symposium on Preventing Major Chemical Accidents, 1987'

'Loss Prevention in the Process Industries' Lees F.P., (2 volumes) Butterworths, 1980.

'Practical Loss Control Leadership', F. Bird and G. Germain, 1957, Institute Publishing, Loganville, Georgia 30249. ISBN 0-88061-054-9.

'Regulating Industrial Risks' H. Otway and M. Peltu, (Science Hazards and Public Protection), Butterworths, 1985.

Index